LANCHESTER LIBRARY
3 8001 00062 4530

KV-057-739

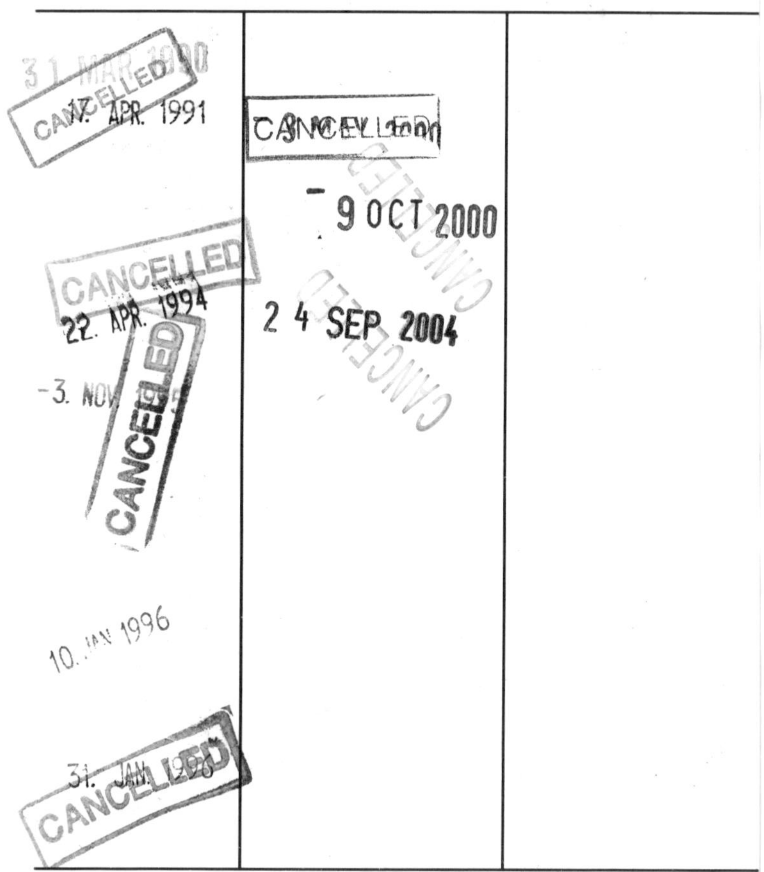
CANCELLED
17. APR. 1991
CANCELLED
- 9 OCT 2000
CANCELLED
22. APR. 1994
2 4 SEP 2004
CANCELLED
-3. NOV
CANCELLED
10. JAN 1996
31. JAN.
CANCELLED

EMPLOYING ROBOTICS IN SMALL MANUFACTURING FIRMS

STRATEGIC IMPLICATIONS

by

J. Michael Alford, Ph.D.
Associate Professor of Business Administration
The Citadel

with Foreword by
Senator Ernest F. Hollings

ROBERT E. KRIEGER PUBLISHING COMPANY
MALABAR, FLORIDA
1988

Original Edition 1988

Printed and Published by
ROBERT E. KRIEGER PUBLISHING COMPANY, INC.
KRIEGER DRIVE
MALABAR, FLORIDA 32950

Printed in the United States of America.

Library of Congress Cataloging-in-Publication Data

Alford, J. Michael.
The strategic implications of employing robotics in small manufacturing firms.

Bibliography: p.
Includes index.
1. Robots, Industrial. I. Title.
TS191.8.A44 1988 629.8'92 87-3017
ISBN 0-89464-222-7
10 9 8 7 6 5 4 3 2

PREFACE

Hundreds of books and articles published in the United States in the past decade deal with productivity problems in manufacturing firms and many have the common theme of identifying how taking advantage of advanced manufacturing technology can improve the firm's competitive position. While this work also addresses that issue, only one manufacturing technology, robotics, was chosen for this study. The results of these research efforts can then be generalized to a broader base and appropriate parallels in other advanced manufacturing technology can be identified if desired. Further, the study was restricted to a somewhat neglected group, the small manufacturing firms. Considering that most of the manufacturing firms in the United States are designated as being in the small business category, it was felt that special attention should be given them because of their actual and potential contribution to the nation's productivity base.

The purpose of this study was to determine strategic implications of employing robotics in small manufacturing firms. Strategic implications were developed from two aspects. First, five case studies were prepared from data collected during onsite visits. The case studies identify the following for each firm: (1) the principal reasons for employing robots, (2) the users' concerns in employing robots, (3) the results of employing robots, (4) the strategic implications of employing robotics at those firms studies, and (5) future plans of the firms insofar as robotics is concerned. A model is presented listing factors to be considered in making the robotics adoption decision. Second, the data collected in the field visits is analyzed and compared with the results of a similar study conducted in Japan. Implications from this study, such as the ability to compete in new markets, are suggested along with recommendations for actions by various external agencies whose activities may impact the decision of the small manufacturer on whether to employ robotics.

The study reveals that the degree of success in implementation of robotics ranged from very successful and profitable implementations to some troublesome situations with less favorable results than anticipated. The study provides evidence that small manufacturing firms can achieve positive results by employing robots in their operations. The analysis of actual versus expected results indicates that typical approaches used in the decision-making process concerning employment of robots focus on individual short-term results and not the synergistic effects that may be obtained through exploiting combinations of the expected results.

The findings of this study are supported by studies of large firms in the United States and of both large and small firms in Japan which employ robots. The parallels in findings of this study and the Japanese study are significant because there are several times as many robots utilized in small firms in Japan as there are in the United States.

For the small manufacturing firm, robotics may play a more significant role in relieving the shortage of skilled workers than in large firms. Although difficulty in financing of automation projects is often given as a reason for the small manufacturer not to pursue automation, this study demonstrates that a number of financing alternatives are available which should be investigated.

FOREWORD

Americans have rediscovered the importance of manufacturing. While this country once led the world in manufacturing, and can do so again, we now know that we let ourselves fall behind. Lagging investment in robotics and other new process technologies has allowed determined and savvy foreign competitors to surpass us in both manufacturing efficiency and quality.

Lack of ingenuity is not our problem. We still lead the world in new scientific discoveries and inventions. But we became complacent about manufacturing. Research in this country focused on new inventions, not improved ways to make them. Others, particularly the Japanese, poured their efforts into improving manufacturing and marketing. As a result, we won the Nobel Prizes and the Japanese won the profits.

Americans now recognize this stiff foreign competition in manufacturing and have begun to respond to the challenge. We are now asking the right questions, about both federal research priorities and, equally important, about how to transfer the technologies our researchers develop to the companies that need them. J. Michael Alford's new book on robotics and small manufacturing firms helps answer these questions and helps point us in the right direction.

Alford examines the benefits of robots in small U.S. manufacturing firms and the barriers now preventing their widespread use. He does so by examining the histories of five firms in the Southeast. As he points out, small manufacturing companies are crucial to the American economy. The United States has over 300,000 manufacturing firms with fewer than 2,000 employees. Other studies have shown that these companies supply 75 percent of the nation's manufactured parts.

Would the sensible introduction of robots help these companies? Absolutely. Alford shows that small and medium-sized manufacturing firms would benefit greatly from automation. They could boost productivity, increase quality, and better compete with foreign companies.

Are America's small firms taking advantage of the new technology? In general, they are not. In 1984, as Alford shows, U.S. firms had installed only 13,000 robots, as compared with 50,000 in Japan. Testimony before the Senate suggests that the gap has still not closed, and other research by Ramchandran Jaikumar of Harvard shows that even those American firms with robots and related flexible manufacturing systems use them less effectively than their Japanese counterparts.

Alford's major contribution is to demonstrate why these small American firms are not taking advantage of the technology that can save them. He examines the standard argument of financial barriers and finds it wanting. Small firms do face capital problems, he says, but a number of financing alternatives are available. Financial planning models do play a critical role, however. Both managers and accountants focus on short-term results, and automation rarely pays for itself in the short run. But robotics, properly used, will yield large dividends in improved quality, new flexibility, and new markets. For that reason, we in Congress support efforts by the Commerce Department and major accounting firms to develop new cost-benefit formulas that include these advantages. These formulas will help address one of the problems Alford identifies.

The most important barrier, Alford suggests, is lack of information. There is a need to educate small manufacturers on the factors to be explored when considering automation. They need information not only on financial aspects but also issues of worker retraining, technology options, and management strategy. Vendors, unfortunately, cannot always provide such information. One of Alford's best insights is that robot producers do not pursue the sale of robots to small manufacturers; queries from small firms often require site visits and engineering studies that are expensive to the vendors and small firms do not understand the need to pay for such services.

The country thus faces a classic chicken-and-egg problem, what the economists call a "market failure." Small firms are understandably reluctant to invest in expensive new equipment until they get the information they need, but large suppliers will find that information too expensive to provide until the market is larger and profitable.

What can be done? Alford's analysis reinforces a direction Congress is already taking. Well-targeted federal programs can play a valuable role here. Just as the agricultural extension program helped make American farming the best in the world, a well-targeted federal technology extension program, working with the states and with our universities, can help provide the information services small manufacturers need. I have recommended the creation of a regional federal-state centers for technology transfer in manufacturing, modelled on an important project in South Carolina which transfers Commerce Department manufacturing expertise and technology to the Navy. That project will cut the manufacture of spare parts for the fleet from 365 days to 30 days. It serves as a role model for efforts to transfer federal expertise already paid for by the taxpayers to Americans who could benefit from it.

As a country, we still have much to learn about modernizing our manufacturing base and responding to the great competitive challenge before us. But industry, labor, the federal government, the states, and academia have finally started to work together. They have begun to make the United States once again the world's leader in manufacturing. J. Michael Alford's analysis is an important step in that long but vital effort.

Senator Ernest F. Hollings
Washington, D.C.

ACKNOWLEDGMENTS

Although I take full responsibility for this study, many deserve recognition for helping in the formulation of the project and urging me to complete it. Several faculty members in the Department of Management at the University of Georgia provided guidance throughout the project. They are Drs. Bill Boulton, John Blackstone, Jim Cox, John Hatfield, and Frank Hoy. Dr. Chuck Hofer, also of the University of Georgia, taught me that there are always more than two sides to any issue. This is a healthy viewpoint for conducting research. Mr. George Munson, a pioneer in robotics, supplied helpful insights when the project was in its early stages. Mr. Mike O'Shaughnessy of Excel 2000 was most helpful in suggesting possible data sources and briefing me on the status of the robotics industry.

The true stars of this story, although they asked to remain anonymous, are the managers and employees of the firms I visited. They donated valuable time and effort which enabled me to collect data. It was gratifying to see Americans hard at work solving problems in a positive and energetic manner.

On a more personal note, thanks are due to my uncle, Colonel Milam R. Smith, for his encouragement and guidance for many years. This work is dedicated to my wife, Anne Trent Alford and our children, Stephanie and Gregg. Individually and together, they kept me on course.

TABLE OF CONTENTS

APPENDIXES

LIST OF TABLES

LIST OF ILLUSTRATIONS

CHAPTER 1

Introduction

Small business has received the attention of researchers in areas of strategic planning, strategic decision making, the impact of government policies and regulations, and to a lesser extent in the technology transfer relating to the small firms. The need for attention to the last topic can be brought into focus by recognizing that firms with fewer than 2,000 employees constitute approximately 99 percent of the manufacturing enterprises in the United States [119]. There have been a number of recommendations to accelerate the automation of manufacturing processes in the United States as a means of improving the competitive position of U.S. firms in the global economy [58, 118]. These recommendations have been directed toward large manufacturers, such as automobile manufacturers, and have neglected the small business sector of the manufacturing industry.

The experiences of U.S. robot users indicate an average improvement in productivity of 20 to 30 percent due to robot employment [129]. These results were based primarily on data collected from large firms in the automotive, aerospace, electronics, and home appliance industries. One study found that only 10 firms accounted for 30 percent of the robots in use in the United States in 1982 [129]. The robot is important to the small manufacturer because it can be a "standalone" element which can interface with existing equipment and human workers without completely automating the manufacturing process. Thus, it may be possible for the small manufacturer to gain some benefits of automation without the prohibitive expense of complete automation.

Because the smaller manufacturing firms are significantly lagging the large firms in the adoption of automated manufacturing technology, it is critical that greater attention be given to smaller manufacturing firms [75]. Ways must be explored to help increase the productivity of this sector. As Donald Koch [71] pointed out, it is time for U.S. manufacturers to consider a "microsolution" to productivity problems. This is one which is centered on the shop floor and proper management strategies. Since management in smaller firms is more directly involved in the activities on the top floor than in large firms, it is evident that the majority of U.S. manufacturing firms stand to benefit from the "microsolution" approach.

One microsolution considered a prime candidate for enabling small manufacturers to improve productivity is the employment of more modern technologies such as robotics. Robotics is seen as a key element in the move toward programmable factory automation [7]. The neglect of studying the strategic implications of employing robotics in the small firm can be attributed to the fact that the U.S. robotics industry is still in its infancy with only 13,000 robots installed as of December 1984 compared with about 50,000 installed in Japan [106] and that of some 50 robot producers in the United States, only 2 had made a profit by the end of 1983 [129]. Profitability in particular has caused the robot producers to seek sales from large firms capable of making multi-unit purchases rather than the one or two unit sale of a small manufacturer. These factors combined to produce a situation in which the number of small firms using robots is very limited. Additionally, the Robot Institute of America reporting procedures do not segregate users of robots in terms of size of the firm. Therefore, an initial effort is needed to determine the strategic implications of employing robots in small manufacturing firms.

Research Objectives

This investigation sought answers to questions relating to the impact of robotics applications in smaller businesses. Specific research questions included:

1. What are the principal reasons for employing robots?
2. What are the user concerns in employing robots?
3. What are the results of employing robots in the small manufacturing firm?
4. What are the strategic implications of using robots in smaller manufacturing operations?
5. What are the future plans for employing robots in those firms studied?

Since the data collected by the Robot Institute of America on the employment of robots is reported on an aggregate basis, there is no way to compare the results of using robots between large versus small firms. Thus it was necessary to contact robot producers and distributors to identify the users who were in the category of having fewer than 2,000 employees. Though most of the over 50 domestic and foreign producers contacted were cooperative, a few considered their customer lists confidential and did not release them. Letters were followed by telephone calls to various persons in the industry and to academic research facilities in an effort to identify users in the desired category.

The research methodology chosen for this study was the development of case studies based on actual visits to user sites. This decision was because of the desire to complete a study which would be practitioner oriented. Actual site visits provided the researcher the opportunity to follow up and obtain immediate clarification of answers provided via a questionnaire. Firsthand observations were also made of the robot applications to supplement the more abstract questionnaire data.

Because of funding limitations, the area of research was restricted to the southeastern United States (Alabama, Georgia, Florida, South Carolina, North Carolina, and Virginia). The combination of correspondence and telephone calls, however, produced only nine sites which would qualify for the study. Five of these agreed to cooperate in the study. Preliminary reports from the research sites indicated mixed results from the employment of robots. Thus it was anticipated that the small manufacturers might have more problems using robots than the large firms. A study of Japanese firms found the employment of robots enabled some of the firms to enter new markets [59]. It was anticipated that such might also be found in U.S. firms.

This study is significant because it addresses an issue that is foremost in the concerns of the United States today which is the competitive position of the United States in the global marketplace. Current sentiment on this issue is found in the following quote.

> In industry after industry, U.S. firms are losing their competitive advantage to foreign firms. . . . The American public is well aware of the threat posed by the industrial strength of certain European countries, and especially of Japan. . . . South Korea, Taiwan, Brazil, Hong Kong, Mexico, and Singapore are now serious challengers to U.S. firms [73, p.1].

At the close of World War II the United States was the world leader, not just as a military power, but also in the production of virtually all forms of manufactured products. Through the 1950s American products dominated markets around the world. U.S. firms produced stylish clothing, advanced electrical appliances, and industrial goods which were demanded worldwide. U.S. productivity was high, plants were relatively modern, and management techniques were considered most advanced. The United States freely shared its products and know-how with other countries without consideration of the potential consequences or even the possibility of role reversal. It was assumed that U.S. dominance would continue indefinitely [17, 73].

Not until the late 1970s did U.S. business and government begin to realize that U.S. leadership had been challenged and displaced by foreign competition in several markets [60, 61]. U.S leadership was no longer a certainty in international dealings. The U.S. productivity had become the lowest in the industrialized world (Figure 1.5). Balance of payments deficits and national debt continued to grow without any sign of reversal through 1984.

These developments had not happened overnight. For example, in 1950, the United States had a 30 percent share of the world export machine tool market. By 1981 the share was down to 10 percent. In fact, in 1981 the United States became a net importer of machine tools [122]. While the trend has continued, it was not until the last decade that the United States began losing share in its strongest markets, such as electronics and automobiles [49, 113]. Figures 1.1, 1.3 and 1.4 indicate the changing U.S. trade situation and Figure 1.2 indicates the U.S. federal budget deficit situation.

It is now recognized that the U.S. firms have failed to reinvest in their manufacturing facilities and technologies, allowing them to age and depreciate. This behavior occurred at the same time that foreign competitors were investing heavily in new production technologies and upgrading their manufacturing skills. Today, for example, executives from the United States go to Japan to observe modern production techniques [73].

The decline in U.S. productivity has been well documented in recent years [9, 17] and the growth rate

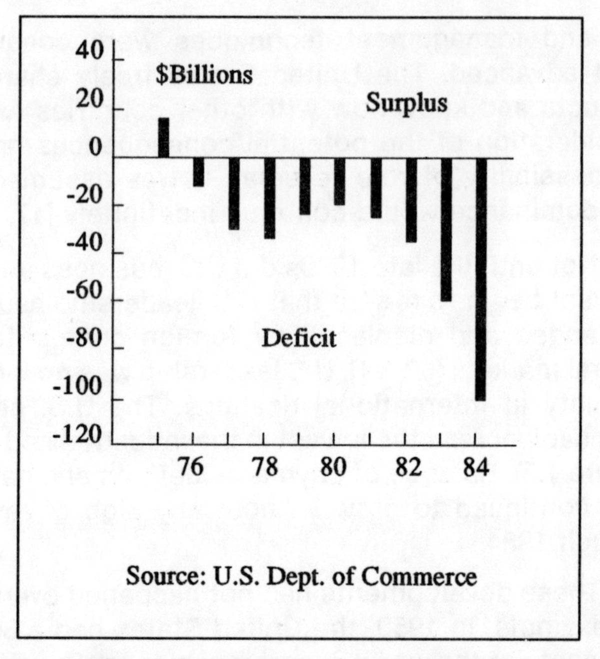

Fig. 1.1 U.S. merchandise trade balance of payments

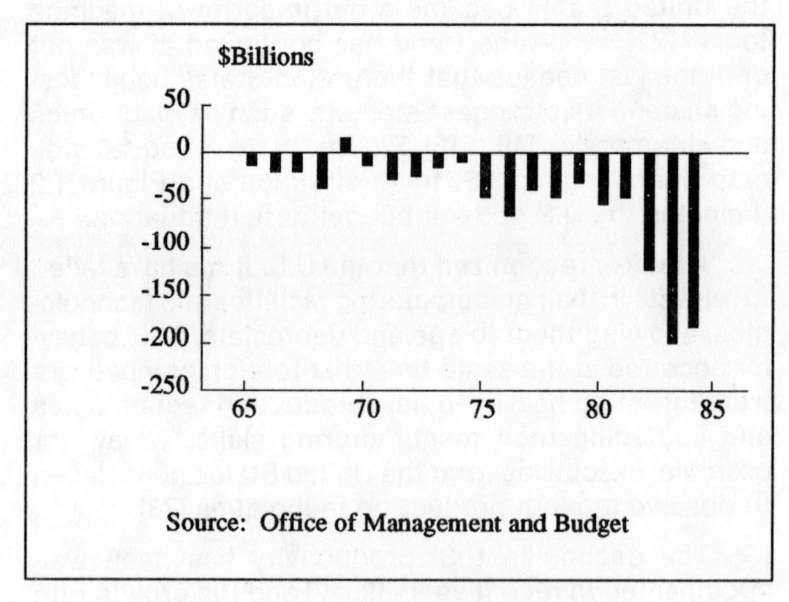

Fig. 1.2 Federal budget deficits

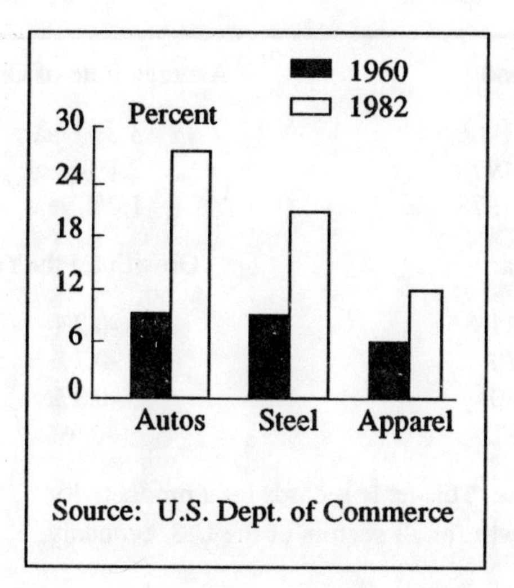

Fig. 1.3 Imports as percent of total sales in the U.S.

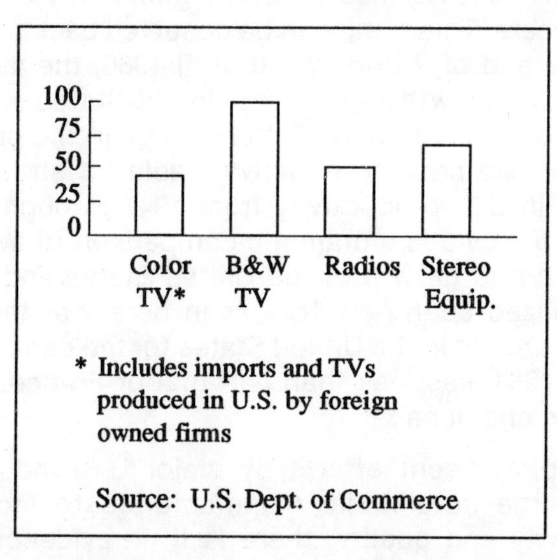

Fig. 1.4 Consumer electronics imports as percent of total sales in U.S.—1982

Period	Average Rate of Growth
1947-1966	3.2%/year
1966-1973	2.1%/year
1973-1978	1.0%/year
Year	**Growth for the Year**
1978	-0.2%
1979	-0.7%
1980	-0.3%
1981	+0.9%

Note: This table records total productivity growth for all sectors of the U.S. economy.

Source: U.S. Department of Labor, Bureau of Labor Statistics

Table 1.1 U.S. Productivity, Output Per Worker-Hour

of U.S. productivity has slowed significantly over the past 10 years. This trend can be observed back to 1966. From the end of World War II until 1966, the average productivity growth rate in the United States was 3.2 percent; by the end of 1977, the change in the productivity rate had become negative. Table 1.1 shows the changes in U.S. productivity from 1947 through 1981. Figure 1.5 provides a dramatic comparison of average manufacturing growth of the United States and other industrialized countries. It is seen here that the productivity growth in the United States for the period 1973 through 1981 was less than half that of France, West Germany, and Japan.

Despite recent efforts by major U.S. industries such as the automobile manufacturers to increase productivity and quality, there is little evidence that these efforts are reversing the trend of continued trade deficits. The U.S. trade deficit for the year 1984 was

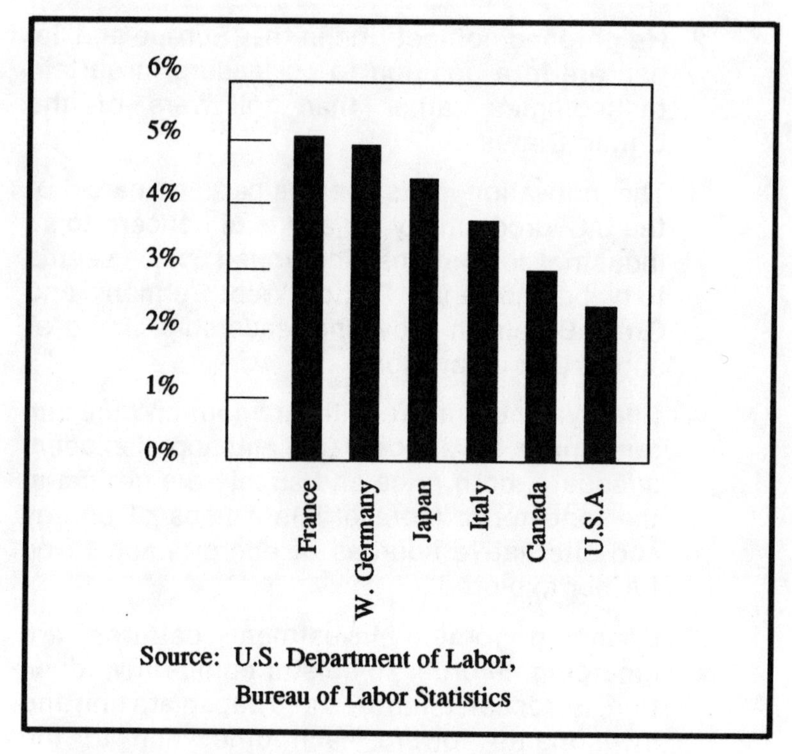

Fig. 1.5 Average manufacturing productivity growth 1973–79

some $101 billion (Figure 1.1). The key to improving productivity according to the Brookings Institute is innovation. Firms investing heavily in developing technology and carrying it forward into commercial products have about twice the productivity rate, three times the growth rate, nine times the employment growth, and one-sixth the price increases as firms with relatively low investment in these activities [82, 93].

Besides the productivity lag, the United States faces global market problems generated by factors such as [47, 66, 67, 81, 83]:

1. Continued protectionism both in the European Economic Community (EEC) and Japan as well as developing nations.

2. Heightened competition in that Europe and Japan are in a position to be leaders in certain technologies rather than followers of the United States.

3. The innovation lag is seen as part and parcel of the U.S. productivity lag and is of concern to all industrialized nations. The United States seems to be behind Japan, France, West Germany, and Great Britain in providing leadership aimed at supporting innovation.

4. Energy concerns are still paramount. While current prices have moderated and supplies seem adequate, both price and supply are critical in the long term. More efficient uses of energy and alternative sources of energy need to be further explored.

5. Changing foreign investment patterns are emerging. In order to bolster competitive position, European countries and Japan are shifting investments toward each other and to the United States. The Nissan plant in Tennessee, proposed Mazda plant in Michigan, and a rapidly increasing number of smaller foreign owned operations in the United States are examples of this. U.S. firms are also investing offshore in attempts to maintain market share and profitability. Some observers see the latter as a move that will ultimately be detrimental to the U.S. position even though it provides short-term gains. The rationale is that the offshore investment causes continued exposure to foreign competitive advantages at the long-term expense of the United States because the investments siphon off funds that should be used to improve U.S. productivity.

6. External debt financing and the significant efforts required to refinance loans to developing

countries are also problem areas. These activities put pressure on obtaining funds by U.S. producers and cause concern about the stability of financial institutions in light of the questionable ability of many third world countries to repay loans.

7. Exchange rate fluctuations with cycles of currency devaluation/revaluation impact as country's export/import balance. Devaluation of a currency does not typically lead to a proportional increase in exports. Yet, a strong U.S. dollar hurts exports while fueling the trade deficits because of increased demand for relatively lower priced foreign goods.

Some call for government intervention, changes in government and the establishment of a formal government industrial policy similar to that of Japan as cures for the present competitive problems of the United States [109, 116]. However, as Beranek and Selby [10] point out, the government view that all firms benefit equally from its program is incorrect. For example, the Economic Recovery Act of 1981 and the establishment of university based innovation centers by the National Science Foundation tend to provide more advantages to large businesses than to the small firms [120].

The determination of the strategic implications of the employment of robots by the small manufacturer is seen as a significant step toward helping solve one of the nation's most vexing problems. Since some 99 percent of the U.S. manufacturing enterprises consist of firms with under 2,000 employees, the most significant improvements in productivity and trade are likely to come from this sector. Yet, little attention has been given to small business productivity.

Overview of the Remaining Chapters

Chapter 2, "Literature Review," presents a review of literature concerned with strategy setting and productivity in smaller U.S. manufacturing firms. This chapter includes additional literature describing the role of robotics in manufacturing, the government's role in technology transfer, and the research methodology chosen for the study, as well as a conceptual framework for the robotic technology adoption decision process.

Chapter 3, "Methodology and Conduct of the Study," focuses on the clinical field research and method of collecting data. A description of the sample used, the research instrument, and methods of analysis are presented.

Chapter 4, "Case Studies," consists of five case studies developed during the research.

Chapter 5, "Analysis of Data." The data collected during the field research is analyzed for the purposes of answering the research questions and is compared with another robotics employment study.

Chapter 6, "Conclusions and Recommendations," presents the final outcomes of the study along with strategic implications and recommendations for furture research.

CHAPTER 2

Literature Review

This chapter reviews the literature pertinent to the employment of robotics with emphasis on those manufacturing firms which have 2,000 or fewer employees. Much of the literature concerning the development and employment of technology applies to both large and small businesses. The review is divided into five parts: (1) literature concerned with strategy setting and productivity in the United States with emphasis on the small manufacturer; (2) literature concerning the role of robotics in manufacturing; (3) literature on the government's role in technology transfer; (4) literature supporting the research methodology selected for this study; and (5) a conceptual framework developed for use in the robotics adoption decision process.

Strategy and Productivity

Strategy has been described as the manner in which a firm achieves a match or "fit" with its environment. There are three levels of strategy: (1) corporate strategy, which involves organizations competing in more than one business, (2) business strategy, which involves competing in a particular industry or product/market segment, and (3) functional strategy, which deals with maximizing the productivity of the firm's resources in support of corporate and/or business strategy [54]. For the small manufacturing firm, the focus is typically on the business level and functional strategies, that is, identifying the product area(s) in which to compete and developing means of being successful in that particular market. The smaller firms can exhibit approaches similar to those of large firms in the competition for business. They can concentrate on a particular market segment and/or differentiate their products from those of the competition through quality, function, and price, or a combination of these factors [97].

The fact that small organizations have received little attention in strategic planning research is well documented [80]. Those investigations that have been made into small business strategic planning practices indicate that formal, systematic planning in the small firm is rare [45, 126]. At the same time, investigations into the value of a strategic plan to the small firm indicate that those with a definite plan of action out perform those firms which do not plan [104]. Of particular interest in the manufacturing sector, as reported by Robinson and Pearce [105], is that those small manufacturing firms that plan are more successful than those which do not.

The need for study of the small manufacturing sector of U.S. industry is evident because in 1979, 69 percent of the manufacturing firms in the United States had revenues of less than $1 million. Further, in 1982 approximately 99 percent of the manufacturing enterprises in the United States had fewer than 2,000 employees [119]. Thus from the standpoint of productivity and global competition, the small manufacturing firms are significant as the United States strives to improve its competitive position.

Although the literature reveals considerable concern over a lack of formal planning, it has been found that many organizations develop effective strategies through informal means [54, 97]. Since many small firm owners fill roles such as chief engineer and production supervisor in addition to being head of the company, a formal written planning system may be too time consuming. Also, there is often a lack of training in planning methods on the part of management. The ready access of the owner to other firm members may be the key to success of the informal systems.

The concept that the structure of a firm should follow its strategy has been discussed for some time [20, 111]. However, this is seemingly more appropriate for the large firm since most of the small firms remain small through their lifetime unless merged with other firms [126]. Carroll's concept of the specialist strategy seems more appropriate for the small firm [18]. This strategy concerns how the firm can best compete in its own area of expertise. For a firm with limited access to resources, it may be more reasonable for the structure and/or technology of the firm to point to the strategy which should be followed. Because a firm maintains a small market share does not necessarily mean that it will have low profitability [76, 138].

The small manufacturing firm which investigates the employment of robots may well find itself in a position to obtain a competitive advantage through the use of a new technology as espoused by Porter [97, 98]. Since investing in technology can be more rewarding than increased investing in labor, the move toward advanced technology may be the better route for the small firm [31, 134]. For example, Engleberger described a scenario in which hourly costs for a robot were only $6 versus $15 per hour for human labor [35].

The relative decline in overall productivity and loss of both domestic and foreign markets to foreign competitors is obvious. The recommended cures for the condition range from noninterference by government to the development of a national industrial policy. There is literature from several disciplines on this matter [30, 37, 53, 64]. The following executive summary from a Robotics Institute of America (RIA) position paper serves to sum up most of the recommendations presented in the literature [106, p. iii]:

> . . . the U.S. is suffering from a negative trend in productivity that can be reversed by:
>
> 1. Encouraging industry and government to invest in applied and new technologies.

2. Setting depreciation schedules that are more compatible with the real work of obsolescence. Grant special tax credits for the added risk involved in accepting and installing new technologies.
3. Changing the patent laws to permit individual innovative entrepreneurs to retain adequate proceeds from their inventive efforts. Transferring dormant technologies from the public sector to useful job-creating products and processes in the private sector.
4. Encouraging individual savings and investment.
5. Encouraging, by tax law changes, the formation of small enterprises based on new research and technology.
6. Encouraging development of meaningful interface mechanisms to permit timely in-flow of new foreign technologies to the U.S.
7. Accelerating the development of government policies, legislation, and regulation designed to encourage more efficient use of our human resources through training, education, and recognition of financial support needed for new and evolving skills and technical careers.
8. Streamlining the myriad components of the Government, Executive Branch, Congress, technical and scientific committee, etc., to work more efficiently to identify goals and promote policies leading to the rapid solution of our productivity problem. Terminate endless, time-consuming and self-serving cosmetic debate, and replace it with action and solution-oriented agreement.

9. Directing top management in acceptance of the role of new technologies and committing their enterprises to these applications and uses.

Legler and Hoy [75] investigated the productivity of small businesses and found that for the period 1965–1976 it remained relatively constant. On the other hand, productivity for the large businesses increased about 14 percent. However, in the manufacturing sector, the small manufacturer had greater gains than the large manufacturer.

One key to improved productivity and a positive move toward regaining a strong competitive position is to adopt programmable automated manufacturing techniques [44, 58]. Robotics is seen as the most viable avenue for smaller manufacturers to become involved in this process. At this time fewer than 20 percent of the U.S. firms buying robots are in the small business category [23]. In contrast, by 1980 an estimated 41 percent of Japanese firms with between 30 and 100 employees were employing robots [23]. Also, the typical small firm in the United States buys only 1 or 2 robots while the large manufacturers tend to purchase from 10 to 200 robots at a time for their automation projects [129]. Thus, the small manufacturers in the United States seem to be moving slowly in their adoption of robotics. Additionally, the producers of robots concentrate on the more lucrative market composed of large firms. About 35 percent of the robots in the United States are used in the automobile manufacturing plants with most of the remainder being in the large aerospace, consumer goods, electronics, and off road vehicle manufacturing industries [129]. Robotics producers estimate that only about 5 percent of the identified robot applications are now in place in the United States. This indicates that many more firms could take advantage of the technology and be in the vanguard of the move toward more productive manufacturing processes. This opportunity should be investigated seriously in light of reported productivity gains of 20 to 30 percent resulting from employing robots [31, 35].

The Role of Robotics in Manufacturing

The purpose of this section is to provide definitions of robotics and identify operating characteristics, roles in manufacturing activities, and the potential benefits to be gained by using robots in small business.

Definitions of Robots

Japan is recognized as the world leader in both the use and the production of robots and has provided some definitions and classifications of them. The Electric Machinery Law of Japan defines an industrial robot as an all-purpose machine equipped with a memory device and a terminal device (for holding things) capable of rotation and of replacing human labor by automatic performance of movements. In Japan robots are classified by method of input information and by teaching method as follows [7, 59]:

1. Manual manipulator—a manipulator that is worked by an operator.
2. Fixed sequence robot—a manipulator which repetitively performs successive steps of a given operation according to a predetermined sequence, condition, and position, and whose set information cannot be easily changed.
3. Variable sequence robot—a manipulator which repetitively performs successive steps of a given operation according to a predetermined sequence, condition, and position, and whose set information can be easily changed.

4. Playback robot—a manipulator which can produce, from memory, operations originally executed under human control. A human operator initially operates the robot in order to input instructions. All the information relevant to the operations (sequence, conditions, and positions) is put into memory. When needed, this information is recalled (or played back, hence, the term "playback" robot) and the operations are repetitively and automatically executed from memory.
5. NC (numerical control) robot—a manipulator that can perform a given task according to the sequence, conditions and position, as commanded via numerical data. The software used for these robots includes punched tapes, cards and digital switches. This type robot has the same type control as an NC machine.
6. Intelligent robot—this robot uses sensory perception (visual and/or tactile) to independently detect changes in the work environment or work condition and, by its own decision-making faculty, proceed with its operation accordingly.

The Robot Institute of America (RIA) defines a robot as [7]:

> A reprogrammable, multifunctional manipulator designed to move materials, parts, tools or specialized devices, through variable programmable motions, for the performance of a variety of tasks.

The RIA definition has become accepted worldwide as the most appropriate for the modern industrial robot. The key words are "reprogrammable" and "multifunctional." Reprogrammable means that if a robot gets a new assignment, it will need a new set of instructions, but its basic structure will not change. (However, in practice, a new end effector or hand is often required for the new task.) Multifunctional means that the robot can perform a variety of tasks. In the ideal situation, the only thing that changes when a new task is assigned to the robot is the instruction program.

The general acceptance of the RIA definition of a modern robot is most important in comparing the number of robots in use in different countries. The RIA definition excludes the first two classes of Japanese robots. Japan has from 85,000 to 100,000 of the simple automatons which fit into classifications 1 and 2 above. Japan uses about 32,000 robots of the high technology type described in the other classes.

The actual number of robots in operation conforming to RIA's definition is uncertain. Over the years, the estimates have varied depending upon the sources of data. The latest data available provides the following [108, 129]:

*Country**	*No of Robots +*
Japan	31,900
United States	7,232
Sweden	2,300 + +
France	993
United Kingdom	977
Italy	600
Others	2,954
Total	46,956 + +

* Excluding Communist countries.
+ Through December 1982.
+ + Estimated.

Significant installations in 1983 and 1984 place the U.S. figure at 13,000 at the end of 1984. The problem with this data is that in some cases it includes unfilled orders. The U.S. robotics industry expects a growth rate of 50 percent in 1985 with a tapering off to around 25 percent a year after 1985 [107]. For growth comparison purposes, Japan had only 13,000 robots (RIA definition) installed at the end of 1980 and the United States had 5,000 [129]. Two years later, Japan had 32,000 and the United States had 7,232 robots.

Operating Characteristics of Robots

Operational classification of robots is by shape and movement since this will determine the application for which they are best suited. Characteristics of robots by shape and by movement are [59]:

> Cylindrical coordinates robots: These robots are usually cylindrical in shape and move up and down and rotate around in the cylindrical coordinates system.
>
> Polar coordinates robots: A very standard type of robot that comes in a number of variations including fixed types (mounted to the floor or ceiling) as well as traveling over the ground or across the ceiling). The polar coordinates robots are applicable to a wide range of tasks.
>
> Cartesian coordinates robots: This type of robot moves primarily along the X, Y, and Z axes, it is particularly suitable for palletizing work. Since its work area is so wide, it can be utilized for handling large-sized work pieces.
>
> Multijointed, articulated robots: This type of robot can be installed in places where space is at a premium and is capable of high speed movement along all axes. It is particularly well suited for installation in a painting booth or when numerous robots have to be positioned at one worksite.

Figure 2-1 illustrates robot motions and movements.

Term	Definition
Cylindrical Coordinates Robots (Figure A)	Manipulators that move primarily within the cylindrical coordinates system.
Polar Coordinates Robots (Figure B)	Manipulators that move primarily in the polar coordinates system.
Cartesian Coordinates Robots (Figure C)	Manipulators that move primarily in the cartesian coordinates system.
Articulated Robots (Figure D)	Manipulators that move primarily by means of multiple joints.

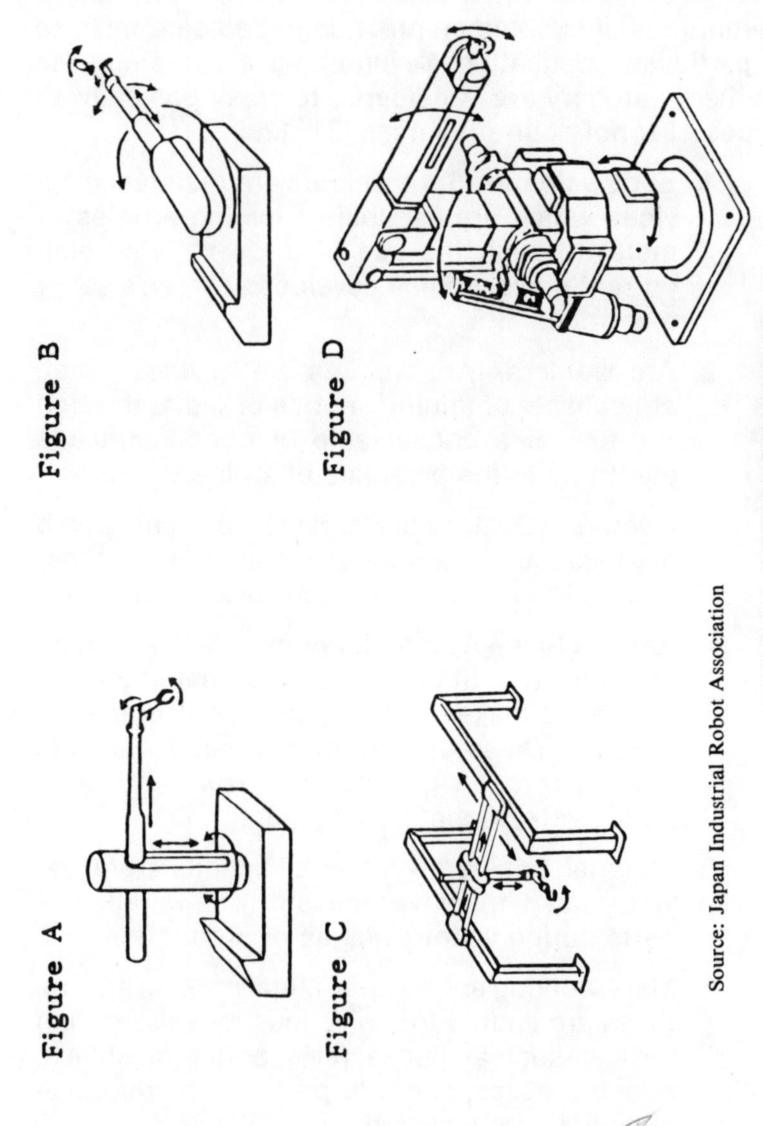

Source: Japan Industrial Robot Association

FIGURE 2.1. FORMS OF MOTION

Applications of Robots

In practice rather than identifying robots by their axes of motion, they are identified by application. Although any given robot could possibly be used for different jobs, it is common practice to dedicate them to a particular application. Seven end-use types and the "other" category are considered to cover basically all types of robots currently used [31, 129]:

1. Spot welders—Spot welders are resistance devices which are capable of joining articles of metal through the use of a low-voltage, high current power source developed across a set of electrodes.
2. Arc welders—Arc welders are devices which are capable of joining articles of metal through the use of a consumable or nonconsumable electrode in the presence of an inert gas.
3. Coaters—Coaters are spraying devices which are capable of applying paint, lacquer, or other liquids to article requiring surface treatment.
4. Assemblers—Assemblers are devices which are utilized to fit or join together manufactured articles to make a subassembly or a completed product. These operations are usually accomplished through the use of screws and nuts, pins, rivets, or similar fasteners.
5. Material handlers—Material handlers are devices used to move and store materials and parts during various stages of production.
6. Metalworking apparatus—Metalworking apparatuses are limited to the various metal-removing devices such as lathes, mills, boring machines, punch presses, and drill presses. Welding machinery is not included.

7. Loaders/unloaders—Loaders/unloaders are used to supply and remove parts or material from other machines (metalworking machines, molding apparatus, and so forth) which perform the manufacturing operation.
8. Other—"Other" includes devices fitting the definition of robots, but not described in 1 through 7 above. Such other devices in types 1 through 7 or other types of robots, e.g., measuring, inspection and testing robots.

Motivations for Employing Robots

There seems to be a consensus across the applications and results literature about the benefits of employing robots whether in the large or small manufacturing firm [38, 132]. The benefits considered in adopting robots are (in rank order) [59]:

1. Reduced Labor costs
2. Elimination of dangerous jobs
3. Increased output rate
4. Improved product quality
5. Increased product flexibility
6. Reduced materials waste
7. Reduced labor turnover
8. Relieving pressure of skilled labor shortage

Although firm size was not differentiated in the above list, the easing of the shortage of skilled labor seems to be more important to the small firm than to the large firm.

Studies on the use of robots in the United States have been oriented toward data collection at large firms. The Japanese have studied both large and small manufacturing firms which use robots. Besides the factors listed above, the Japanese found the following as the result of employing robots [59]:

1. Improved equipment investments—shortened time to realize returns on investments.
2. Decrease in labor accidents.
3. Made production control more efficient—more consistent quality meant less stock on hand, which in turn reduced inventory costs, enabling production plans to be more readily prepared.

Aside from the general findings above, there were some findings peculiar to small manufacturers in both motivation and results. These were:

Motivations:

1. Enterprising management—the president of the firm adopted the new technology early.
2. Direct and indirect influence of major customers. Direct instructions from or pressure in terms of increased orders from major customers who were using robots.
3. Too many older workers and not enough skilled labor available to the small firm.
4. New materials and processes make automation more practical.

Results (based on ten small firms) [59]:

1. Improved working conditions
2. Enhanced productivity
3. Improved product quality (precision)
4. Labor savings
5. Reduced labor costs
6. Improved worker attitudes and capabilities
7. Stabilized production and readily adjustable output
8. Better able meet delivery dates
9. Improved safety, fewer accidents

10. Solving of skilled labor shortage problems
11. Capability to deal with specification changes
12. Increased product reliability, fewer inspections
13. Product diversification and expanded business
14. Heightened trust on the part of customers, resulting in more orders
15. Reduced equipment costs
16. Reduced stock on hand

The results of using robots meant that some small firms were able to attract qualified technicians and stabilize labor problems.

Cost of Robots

Because of the many types of robots and the number of manufacturers (some 50 firms in the U.S., 250 Japan, about 20 in Europe), the costs of robots differ greatly [96, 134]. Robots, like some other systems, can be bought with control systems of varying sophistication and other interface and support equipment. In some cases, the "robot" part of the system ranges from 20 to 50 percent of the overall cost of an installed robot system. For example, in a $35,000 robot installation at Company A, a small manufacturing firm in Mississippi, the robot cost $16,000, control and support equipment plus engineering services and installation cost $19,000. A recent sample of robot cost data is presented in Table 2-1.

RIA reports that the average price of robots in 1983 was $74,100 and that by mid-1984, the price was down to $59,100. Prices were expected to remain steady throughout 1985 at about the same level as in 1984 [107].

Table 2.1 Robot costs: Range and average prices of foreign and domestic robots, 1983.[1]

(Thousands of Dollars)

Type of Robot	Domestic				Foreign			
	Low	Med	High	Ave	Low	Med	High	Ave
Spot welders	65	83	160	78	50	86	112	87
Arc welders	33	95	100	90	29	73	73	37
Coaters	20	103	170	138	60	90	197	80
Assemblers	43	62	250	106	12	50	77	40
Material handlers	20	70	150	78	38	58	104	66
Metalworking apparatus	47	140	233	154	73	120	na	85
Loaders/unloaders	16	71	131	86	38	75	112	70
All others	47	86	175	64	25	64	109	64

[1]Prices are net delivered prices (less discounts and allowances) of the robots delivered to purchasers' U.S. facilities.
Source: USITC Publication 1475, December 1983.

The Impact of Robots on the Workforce

As mentioned above, the Japanese small businesses have seen an improvement in their labor situation as a result of employing robots. In the United States the little data available on the impacts of robotics is from the large manufacturing firms and it is difficult to separate data relating only to robots from other advanced manufacturing technologies. However, it has been estimated by Ayres and Miller [7] that robots might have displaced between 8,000 and 12,000 people by 1980. An estimate of 10,000 jobs lost would represent one-seventh of 1 percent of the approximately 8 million semiskilled workers in the manufacturing industry. This is felt to be a negligible impact. There have been concerns expressed that the concentration of robots in the primary metals, fabricated metal products, machinery, transportation, and in precision instruments could lead to pockets of unemployment in the geographic areas where these industries are concentrated. Historically, there have been concerns about the impact new technologies would have on employment and each new technology serves to keep those concerns alive [3, 43, 51, 135]. However, since present methods of reporting data on the employment of robots is not segmented by industry or by firm size, the exact impact of robotics is not determinable, especially for small manufacturers.

The Government Role in the Transfer of Robotics

That the government has a formal role in the formation and operation of private business is well documented [64, 128, 130]. Some government activities affect large and small firms alike. Other actions and programs specifically target the small business sector as is evidenced by the 1953 founding of the Small Business Administration. Government is involved in commercialization of processes and products from government funded research and development. It concerns technology transfer and is a buyer of goods and services produced by the private sector. The traditional policy areas of government involvement with private business lie in the areas of regulation, taxes, and patents. During the past two decades, government has been more active in the areas of personnel policy, acquisition and assistance, and the mandating of technology. The myriad of government involvements does not constitute a comprehensive, coordinated effort, but is the result of piecemeal effort by the various agencies. The impacts are felt on virtually all firms. The small business with fewer than 25 employees or not over $50,000 in government contracts is exempt from some governmental reporting requirements. Although the primary emphasis here is with activity between the federal government and business, it must be remembered that the state and municipal agencies have a similar interface with business. Other areas receiving current attention are those government/university and industry/university linkages designed to promote the advancement of innovative efforts. Although the literature supports the thesis that small firms are heavily involved in innovation, productivity improvement, and technology, it is reported that the university and government programs tend to join with large highly visible firms [11, 63]. Further, as Soergel points out, the government program requirements discriminate against the small business and the independent innovator [120].

Regulation

The two distinct groups of regulations that affect private enterprises are: (1) environmental, health, and safety (EHS) regulations, which are enforced by OSHA and EPA and (2) economic regulation in the energy, communications, and transportation sectors. There has been significant deregulation in the latter three areas in recent years as the Reagan administration has slowed or ended certain regulatory activities [73, 128].

Most literature on this subject deals with the impact on innovation of regulatory strategies aimed at setting standards or actually mandating technologies to be used to comply with standards. Apparently, the issue has been resolved in favor of standard setting, thus allowing firms some flexibility in meeting the standards. The hope is that innovation and cheaper approaches to compliance will result. Areas now covered under the direct standard method are: (1) air and water pollution, (2) drug regulation, (3) workplace safety, (4) registration of pesticides, and (5) identification of toxic substances [128]. Robotics is seen as a way for small manufacturers to use technology to comply with regulatory standards. Robots can work in a number of areas which are hazardous to humans because of toxic substances, heat, and other worker safety aspects. To pursue robotics as a means toward this end may relieve the operator of either a small or a large firm from a costly regulatory burden and enable investment in capital equipment. Estimates are that each dollar spent on pollution abatement equipment reduces spending on productive plant and equipment by some 33 to 40 percent [28]. In regulated industries, it was found that financial reporting requirements have had a negative influence on research and development spending in small firms [57].

A complaint by small businesses about government rules, other than the high costs of compliance with EHS requirements, lies in the regulatory restrictions on cooperative research efforts and export trade regulations. There is a continuing call from the manufacturing industries for protection from imports as well [106]. The Reagan administration maintains that protectionist legislation would ultimately make U.S. firms less competitive in the world market. Historically, because of legal restrictions and the complication of international tariff systems, the small business portion of export trade has been small [22, 33]. For example, in 1978 only 30,000 out of 376,000 manufacturing companies were involved in export. While nearly 70 percent of those 30,000 firms had 2,000 or fewer employees they accounted for under 40 percent of manufactured exports. In contrast, 50 large firms accounted for some 34 percent of manufactured exports [29, 119].

On October 8, 1982, President Reagan signed the Export Trading Company Act. The purpose of the act was to stimulate Export Trading Company activity by providing two primary incentives: (1) institution of a new form of antitrust protection for joint exporting and (2) permission for banking houses to own an interest in the export combinations [119]. Ideally this would prove more attractive to small firms. Additionally, commencing with FY 1984, the Eximbank agreed to set aside funds to finance exports by U.S. small business. While aimed at the improvement of the overall U.S. position in international markets, it was hoped that small firms would take advantage of the act, upgrading existing manufacturing technology through investments in robotics or similar technology in order to improve productivity and the quality of products [119].

Tax Policy

A firm may use resources for capital investment for the purpose of upgrading its own technology or for obtaining the technology developed by others. Tax policies can affect the method selected. For firms with a wide range of research and development possibilities or investment options, the tax structure can impact where emphasis is placed. A relationship is presumed to exist between research and development efforts and gains in innovation [24]. In the 1954 tax code revisions, firms could treat research and development costs that were not in plant and equipment as current costs in order to reduce the immediate tax burden. While this would be more advantageous to the large firm, concessions to the small businesses were made over the past 30 years in the form of various tax code revisions [119]. This could be interpreted as an inducement for the small firm to invest in innovative technology such as robotics.

Studies by Fromm [42] and by Visscher [133] have examined the connection between the level of investment and changes in sales, the investment tax credit, liberal use of accelerated depreciation, and other special tax incentives. The strongest link seemed to be between sales and investment. This has led to a number of recommendations that federal tax policy be oriented toward maintaining aggregate demand and stable prices [93].

Tax proposals designed to stimulate innovation usually fall into four categories: (1) special tax benefits for research and development expenditures (investment tax credit), (2) larger allowable deductions for support of research work in universities, (3) lower tax rates for business income from successful innovations, and (4) expanded opportunities to use business losses from technology-based enterprises as taxable income offsets, either against successful enterprises under the same ownership, or against future income from the same enterprise [24, 128].

Small business spokespersons have soundly criticized the Reagan administration's tax programs such as the Accelerated Cost Recovery System (ACRS) as being primarily beneficial to large firms. The National Small Business Association claims that nearly 80 percent of the savings under ACRS will go to the top 1700 firms [33]. The current stand by small business organizations such as the National Small Business Association, the Mid Continent Independent Small Business Association, and Small Business United are critical of Reagan's tax policies and the failure of the U.S. Chamber of Commerce and the National Federation of Independent Businesses to take a more firm stand against those policies [33]. It appears that the problem is that tax programs like other government programs are devised under the assumption that the impacts will be the same on all firms regardless of their size.

Patent Policy

The government's role in patent policy and technological innovation can be addressed in three general questions [123]:

1. Does the granting of exclusive rights to inventors promote utilization of government sponsored technological innovation better than acquisition of title by government?
2. Do patents in general act as incentives to technological innovation?
3. Does the currently available patent data support the contention that the United States is "losing its genius" for technological innovation when compared to other developed nations?

In July of 1981, the Bayh-Dole Bill introduced changes to the title policy and license policy of the patent laws. These changes favored small business in the obtaining of patent rights from federally-funded research [26, 55, 128]. Kitti [69] summarizes research on the incentive value of patents as supporting the argument that longer patent life is likely to lead to both higher levels and earlier introduction of innovation. On the other hand, Kitti found that a major hurdle to innovation in the patent policies is the patent litigation where about one-half of the challenges are successful. The cost of litigation in these cases is seen as a deterrent to the small firm because of limited resources for the legal services required. Other factors cited are those such as safety testing which is required after the granting of a patent. The delays in getting products to the market thereby shortens the life of the patent.

The third concern stems from an increasing number of patents granted by the United States to foreign inventors. The number of patents in this area increased 91 percent between 1966 and 1975. In 1975, 35 percent of U.S. patents granted were to foreign residents. Since 1970, Japanese patenting in the United States has increased over 100 percent in almost every major industrial category. A review of patents granted in robotics revealed a heavy Japanese and European input compared with patents granted to U.S. residents. Schiffel and Kitti maintained that an analysis of patent data does support the idea that rising levels of exports and economic interdependence explain a good deal of the increase in patent filings by foreign inventors [114]. If this is true, then the lack of export activity by small manufacturers in the United States could be considered as an impediment to innovation and productivity.

Recent Calls for Practitioner Oriented Research

As Thomas and Tymon report, there have been many criticisms that emphasis on rigorous quantitative research methods in organizational science has directed energy away from the considerations of the relevance and the practical utilization of research findings [125]. Thomas and Tymon developed a research model around five key needs which should be considered in making research useful to the practitioner. These are:

1. Description relevance—accuracy of the research findings in capturing phenomena encountered by the practitioner in the organizational setting.
2. Goal relevance—correspondence of the outcome (dependent) variables in a theory to the things the practitioner wishes to influence.
3. Operational validity—ability of the practitioner to implement action implications of a theory by

manipulating its casual (independent) variables.

4. Nonobviousness—the degree to which a theory meets or exceeds the complexity of common sense theory already used by a practitioner.
5. Timeliness—concerns the requirement that a theory be available to practitioners in time to use it.

Fredrickson [39] states that the value of strategic process research will be realized only when it provides an understanding of important phenomena and helps improve organizational performance. Daft in an essay on organizational research calls for learning about organizations firsthand by visiting them [27]. Morey and Luthans call for expanding the concepts of research to include subjective/qualitative methodologies [90]. Finally, Shrivastava and Mitroff blame the limited use of organizational research for practical decision making on the failure of researchers to consider the "frame of reference" of the organizational decision maker [117].

This study is aimed at identifying the practical concerns of managers in the employment of robots in manufacturing processes and determining the strategic implications of adopting robots for the small manufacturing firm. Thus, it should result in a set of recommendations that will be of immediate value to the practitioner as well as serve as a basis for further research.

Conceptual Framework

Technology is only one of several factors which shape the strategic environment of a firm. Although the emphasis in this study is on a specific manufacturing technology, the other factors must also be considered in the development of a decision-making framework regarding the adoption of a new technology. Some important issues affecting the competitive environment are shown in Table 2.2.

Table 2.2 Important Factors in the Competitive Environment

Social-cultural	*Economic*	*Technological*	*Political/Legal*
Social Expectations	GNP Trends	Federal R&D Spending	Regulatory Controls
Social Norms	Interest Rates		Tax Laws
Consumer Activism	Money Supply	Industry R&D Spending	Special Incentives
Demographics	Devaluation/ Reevaluation	Patent Protection	Foreign Trade Laws
Change in Workforce Structure	Energy Situation	New Products	National Economic Policies
	Inflation Rates	Technology Transfer	Anti-trust Laws
	Foreign Competition	New Technological Processes	
	Global Interdependence		

Source: Adapted from W. R. Boulton, *Business Policy: The Art of Strategic Management*, Macmillan Publishing Company, New York, 1984

As Porter states, the major competitive forces in a given industry are [97]:
1. Threat of new entrants into the industry
2. Rivalry among existing firms
3. Threat of Substitute Products or Services
4. Bargaining Power of Buyers
5. Bargaining Power of Suppliers

As Porter states, the major competitive forces in a given industry are [97]:

1. Threat of new entrants into the industry
2. Rivalry among existing firms
3. Threat of Substitute Products or Services
4. Bargaining Power of Buyers
5. Bargaining Power of Suppliers.

These forces impact firms of all sizes and in all industries. Past fluctuations in energy costs and court rulings have shown that even regulated industries are subject to external forces.

Strategy is generally considered to be set at three levels: corporate, business, and functional.

- Corporate Level Strategy (Multibusiness)
- Business Level Strategy (Single Business)
- Functional Level Strategy (Support Business Level Strategy)

At the functional level, the primary purpose is to maximize productivity within the constraints of the business level strategy [54].

In a firm which is in a single business, the corporate level and business level strategies are considered to be the same. Moreover, in the smaller manufacturing firm, the line of demarcation between business and functional or operational strategies becomes indistinct. It seems that the technology of the production processes set the tone for business strategy rather than finding the operations strategy as a distinct subset of the business strategy. The CEO's heavy involvement in day-to-day operations and the informal strategy setting policy contribute to this phenomenon [25, 80, 126].

In 1981, Wheelwright reported on the strategic significance of operations in Japan [136]. He described

In Japanese companies **Strategic operations policy view**	**In U.S. companies** **Traditional view**
Strategic	Strategic
Capacity	Capacity
Facilities	Facilities
Vertical integration	Vertical integration
Production processes	Production processes
Work force	
Quality	
Production planning and control	
Operational	Operational
Implementation and execution of supporting operating tactics	Work force
	Quality
	Production planning and control
	Implementation and execution of supporting operating tactics

Source: Wheelwright, S.C., *Harvard Business Review,* July-August, 1981, p. 72.

Fig. 2.2. Manufacturing Decision Categories Considered Strategic.

what is termed a commitment to a "strategic operations policy" on the part of the Japanese. Figure 2.2 depicts the differences between the Japanese approval and the U.S. approval to strategy.

It appears that, whether by design or happenstance, in the normal course of operations the small manufacturing firm in the United States is closer to Wheelwright's model describing the Japanese companies than to the typical larger U.S. company. While this background provides a brief look at how the small manufacturing firms may not fit the typical strategy models which were developed primarily from studies of large firms, it also indicates a need to develop a more specific decision framework that can be used to evaluate the strategic implications of the adoption of robotics technology by the small manufacturing firm.

Decision Framework for Adopting Robotics Technology

Capital investments by manufacturers are typically based on the economic factor analysis of criteria such as return on investment or payback period or some more sophisticated analysis of the economics of the investment. Some noneconomic considerations such as increased productivity, decreased scrap, and less rework are converted into dollar figures for the analysis as well. It is recognized that a variety of economic analysis techniques can be used. It is also obvious that economics is important to a firm considering the employment of robotics. However, the economic analysis is only a part of the overall picture to be considered by the potential user of robotics in making the decision to adopt or not to adopt robotics technology.

Fleck has identified other factors including technical, managerial and organizational, and labor in a two year study of 32 firms adopting robots in the United Kingdom [38]. Fleck also found that firms with previous automation experience and those initially employing robots in jobs undesirable for humans were more successful in adopting robots. John P. Van Blois of IBM calls for justifying robotics via economic models, especially the DuPont Asset Management Model [132]. However, more important is Van Blois's recognition of the need for a strategic justification which considers technology advancement, competitive position, investment-capital growth, and modernization along with the more common short-term or tactical justification based primarily on payback periods, productivity increases, quality, scrap, rework, and reaction to product changes. Van Blois is especially critical of the typical direct-labor cost versus robot cost per hour in a one-to-one comparison. However, data such as a $15 per hour human labor cost versus $6 per hour total robot cost acts as a very powerful inducement to management [35].

P. J. Rosato of Unimation calls for involving each department in the organization through assignment of specific responsibilities during a four phase robot implementation [110]. The phases are:

1. Initial Robotic/Automation Education
2. Prepurchase Investigation/Education
3. Purchase
4. Training, Installation, and Operation and Maintenance

While Rosato provides a worthwhile approach, especially recommendations concerning management activities in the human resources management and operations/maintenance areas, he has not provided any specific costing or economic analysis approaches or models. Gustafson, on the other hand, supplies a method of choosing a system based on lowest unit cost for a given production volume [50]. This analysis could indicate choosing a manual, fixed automation, or flexible automation production system. The shortcomings here include a lack of long-term or strategic considerations and the other noneconomic factors mentioned thus far. Abair and Logan of Prab Robots, Inc. propose an eight-step process for robotics implementation [1].

1. Basic Research of Available Robot Technology
2. Application Identification
3. Vendor Considerations
4. Application Documentation
5. Selection
6. Design and Build

7. Plant Site Preparation and Installation
8. Training and Spare Parts

In this instance, the orientation is primarily toward technical consideration of the systems.

There are certain technical considerations worthy of mention in robotics adoption. Such factors should be identified by the robotics vendor. However, the purchaser should be aware that while installing new manufacturing technology may require changes in parts routing and floor layout, in some cases the actual design of assembly parts and components may require changes. This is because of limitations of the robot sensing and gripping devices. Identifying and overcoming these obstacles can be time consuming and involve outside suppliers in some instances. J. R. Bailey of IBM provides a number of examples of preferred shapes and/or sizes of assembly parts to be handled by robots [8]. Bailey's guidelines call for designing to minimize numbers of parts and assembly directions. Component design should assure smooth parts feeding, self-aligning and fixturing, and chamfers to accommodate positioned uncertainty. Common interfaces and one-handed adjustments are also recommended. This factor is brought out because of its criticality and it also presents an alternative in deciding on the type and capabilities of the end devices on the robot arm. There are trade-offs in the cost of more sophisticated robotics versus component redesign. Redesign of parts to accommodate robots instead of designing robot devices to fit the user's particular parts can lead to standardization in robotics and parts design. Of particular importance to the user is the opportunity to improve standardization, reliability, and serviceability of the end products as well as the savings in the cost of the robot.

Other factors potential users of robotics should consider relate to aspects of the selection of vendor.

There are about 50 producers in the United States and there are a number of foreign models available. This means that the user has the opportunity in some applications to seek bids from several sources. Six of the U.S. firms account for about 60 percent of the U.S. market share and some of these have heavy industries such as automotive manufacturing as prime customers [13]. Some major producers, however, have from the outset provided products for applications at the lower end of the market in terms of size and/or number of robots per installation. Basic considerations in selection of a vendor include user requirements for engineering and installation services, training, maintenance and spares support, overhaul, and degree of integration into automated systems. There is also the alternative of leasing robots. Prab Robots, Inc. leases robots for as little as $1,000 per month on a three-year lease [13]. Other firms offer similar services.

The purpose of mentioning such wide-ranging factors in robotics adoption thus far is to illustrate the need for considerations beyond the typical economic analysis of equipment costs and payback based only on direct labor savings when determining whether to adopt robotics. Additionally, the proliferation of publications in the field tends to reflect the particular background and/or interests of the individual writers which results in emphasis of some features and neglect of others. The limited background presented here also provides a basis for moving toward a framework of decision factors in the adoption of robotics.

Among the approaches possible, one might consider the following as major divisions: strategic and tactical factors, economic and noneconomic factors, factors related to organizational structure such as technical, human resources management, and financial, or a combination of these approaches. The strategic and tactical approach shown in Table 2.3 is more encompassing; hence it is used here. Strategic factors

Table 2.3 Decision Factors in Robotics Adoption

STRATEGIC FACTORS	*TACTICAL FACTORS*
1. Capital Investments	1. Technological Change
2. Competitive Position	2. Economic Analysis/ Cost Structure
3. Technological Change	3. Vendor
4. Industry Structure	4. Work Force
5. Product/Market Growth Stage	5. Training
6. Organizational Structure	6. Productivity Increase
7. Economic Analysis	7. Quality, Scrap, Rework
8. Political-Legal Factors	8. Parts/Component Design
9. Demographics	9. Inventory
10. Government Policy	10. React to Product Changes
11. Other Factors as Applicable	11. Government Policy
	12. Other Factors as Applicable

typically represent long-term considerations, whereas the tactical factors represent those of a more immediate or operational nature.

The following questions are typical of those to be addressed in the robotics adoption decision process.

Strategic Factors

1. Capital Investment
   - What will be the long-term effect of the investment?
   - Is this a step toward a flexible manufacturing system or a stand-alone installation?
2. Competitive Position
   - Is the investment required to maintain competitive position?

- Will the investment in the technology improve competitive position in the long term?

3. Technological Change

- Is the technology required to maintain or improve productivity, and/or quality over the long term?
- What other benefits will accrue from the changes in technology?
- Is this decision an incremental or a significant change in the technology employed?
- Can the technology be readily upgraded?
- Is the technology similar to or radically different from that currently employed?
- In either case is the technical ability available to effectively use the technology or can the expertise be acquired?
- Will the technology improve process design, product design, quality, all of these?

4. Industry Structure

- If this technology is widely adopted how will it affect industry structure?
- Will low cost technology invite new competition and/or increased competition among existing firms?
- Will high cost technology affordable only by a few of the present competitors, result in a decline in the number of firms in the industry?
- Will fragmented industry move toward concentration as marginal producers who do not adopt the technology fail?
- Will technology desirable to powerful buyers and/or suppliers result in vertical integration?
- Can the potential players and likely survivors be identified and what is likely to be the relative positioning in the market?

5. Product/Market Growth Stage
   - Must mature product/market update technology or exit the industry?
   - Will early stage adoption of the technology provide a significant advantage?
   - Will mature stage adoption of the technology provide product differentiation to revive market growth?
   - Will technology provide ability to quickly change product design to meet shifting demand and competitors' moves?

6. Organizational Structure
   - How will the adoption of the new technology change the basic organizational processes such as communications flows, lines of authority, accounting procedures, departmentation?
   - Will adoption of the technology cause curtailment of operations in some areas (plants) and/or increase activities in others?
   - Will new management systems be required? At what levels of the organization? How long will it take to put them in place? Is the requisite managerial talent currently available or can it be acquired?
   - What sources of potential resistance to change exist? How can they be overcome?

7. Economic Analysis
   - Questions in this area should address capital constraints because adopting robotics may require significant changes in physical plant layout and organization structure which may drive future strategies of the firm. Also see comments in this area under Tactical Factors below.

8. Political-Legal Factors

   - What are the impacts of the technology on:
     employment agreements in force?*
     OSHA requirements?
     commitments to community?

9. Demographics

   - Will the technology decrease the number of workers needed in the long term?
   - Will projected work force membership provide adequate skills and numbers of people with these skills?
   - Must long-term internal training programs be established or will outside sources be adequate?
   - How will shifting population patterns affect decisions on facility closings and planning for new facilities?

10. Government Policy

    - What is the impact of the current tax structure?
    - What is the impact of current depreciation schedules?
    - Are there government programs available to assist in technology upgrades in terms of grants, guaranteed loans, etc.?
    - Is this an industry affected by imports; will government policy impact competition in this arena?

11. Other Factors as Applicable

    - Various firms will have different views on which are the most important considera-

---

*The provisions of employment agreements vary, but typically include advance notice provisions, retraining, job security, etc. For more detail see Robert U. Ayres et al., *Robotics: Applications and Social Implications*, Ballinger Publishing Company, Cambridge, MA, 1983, Ch. 5.

tions. The key question to investigate here is, "Have all factors which will impact the decision been considered?"

Tactical Factors

These factors deal primarily with the near term. There is some commonality between the strategic and tactical factor considerations.

1. Technological Change
   - How long from decision date will it take to have the new technology on line and operating as expected?
   - How will current operations be impacted during the changeover?
2. Economic Analysis/Cost Structure
   - This is an area of both strategic and tactical importance since decisions made impact both the long-term and near-term future of the organization. Additionally, several of the tactical factors mentioned below play either a direct or indirect role in economic analysis because they represent identifiable expenditures and potential cost savings both in the near-term and the long-term. Topperwein et al. furnish some examples of cost savings in motor vehicle manufacturing activities. Most vendors should be able to provide potential users with data on their experiences in actual working systems or laboratory data or similar experiences with equivalent systems of other producers. Topperewin et al. also supply a basic economic analysis model for determining payback period and ROI [127]. Factors included for analysis are: robot cost, accessories costs, related expense, engineering costs, installation costs, tooling

costs, direct labor savings, indirect labor savings, maintenance savings, other costs, other savings, and depreciation.

- When using a model developed by someone else or developing one's own model, it is advisable to examine the definitions which accompany each of the economic factors to ensure that all those things appropriate for the analysis are included.
- Of course, the alternative of leasing should be investigated. It could be particularly beneficial to those who have a strategy of incremental progress toward an integrated system. This could allow for equipment upgrade as the project progresses if more suitable equipment becomes available. The firm's capital structure and equipment requirements could also dictate leasing.

3. Vendor

- What services are available from the vendor (whether producer, systems house or distributor) and which are required in terms of:
  Appropriate technology/equipment
  Engineering services
  Site preparation
  Installation
  Training
  Maintenance support
  Spare parts support
  Financing.
- There are many specific factors to be considered in this area. For example, electrical power capacity and outlet locations, climatic control, interface systems, and so forth which require exchange of considerable information between user and vendor during the bidding stages and all the way to placing the system in operation plus follow-

up. Abair and Logan [1] recommend using a project approach. This certainly is advisable for a major robotics installation, but may be cumbersome for those installing only one or two robots. However, the basic factors should be considered in any case.

4. Work Force
   - What is the immediate impact on direct labor and supervision in terms of skill mix and number of employees?
   - What is the effect of current labor agreements?
   - What actions must be taken in these areas?
5. Training
   - What type of training is required?
   - How long will training take?
   - Will training be conducted in-house or elsewhere?
   - Are qualified trainees available in-house?
   - Should there be a transition from vendor training to in-house training capabilities? Is there staff for this?
6. Productivity Increase
   - How do manufacturers' claims about productivity increase compare to actual user experience?
   - How long will it take to achieve the expected increase in productivity?
   - Will equipment reliability support this expectation?
7. Quality, Scrap, Rework
   - Is there evidence to support vendors' claims in this area?
   - Can the evidence be verified?

8. Parts/Component Design
    - Must assembly parts/components be redesigned to accommodate the new technology?
    - At what cost?
    - How long will it take?
    - Who will be involved, i.e., outside suppliers or in-house?
    - What are the trade-offs between component redesign and robotics configurations?
    - Will advantages such as simplified design, improved quality, and flexibility be gained?
9. Inventory
    - How will the new technology impact raw materials, work in process, and finished goods inventories?
    - Are there potential savings here?
10. React to Product Changes
    - Will the reprogrammable features of the robotics allow alterations to product design/function to improve quality or to react to competitors' moves?
    - Can the same robotics be used to produce a new line of products?
11. Government Policy
    - What government policies and/or regulations will be applicable at the federal, state, and local level?
    - What about tax incentives?
    - Are government grants or loan guarantees available?
    - Can the desired technology be purchased through Industrial Revenue Bonds?
    - Is job training program funding available?

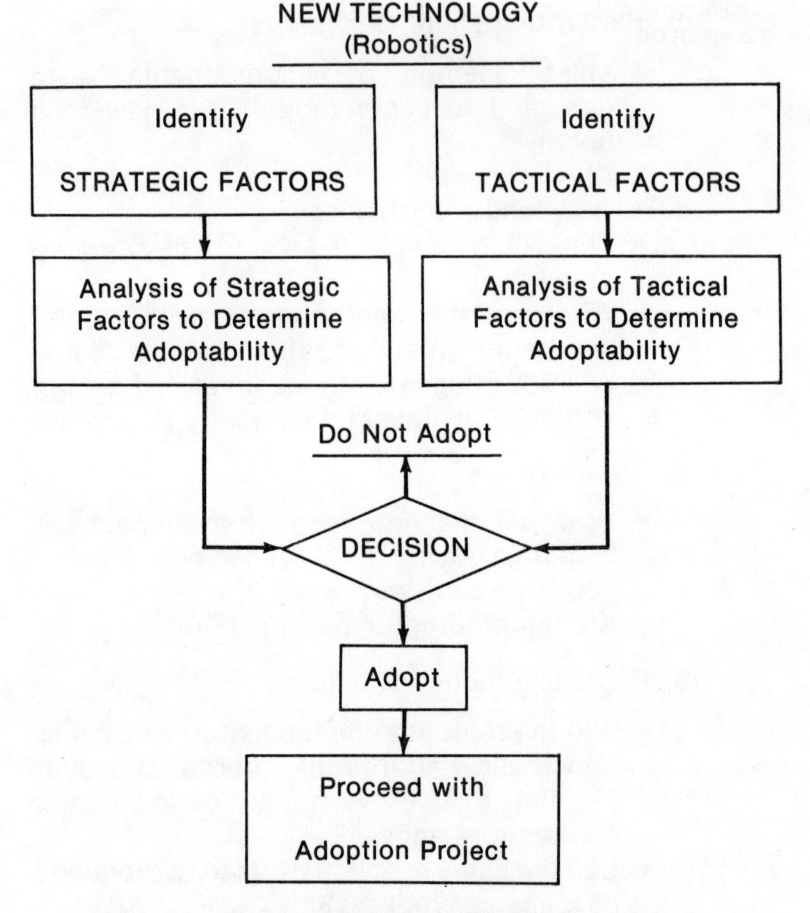

Fig. 2.3. Decision Model for Robotics Adoption.

12. Other Factors as Applicable
    - What factors were not identified in the list above?
    - Have inputs been invited from all personnel?

The framework described above is represented diagrammatically in Figure 2.3. Robotics technology has the potential to impact every department of an organization. Figure 2.4 identifies some areas and types of impact that could be expected from employing robots. The research instrument found in the Appendix was

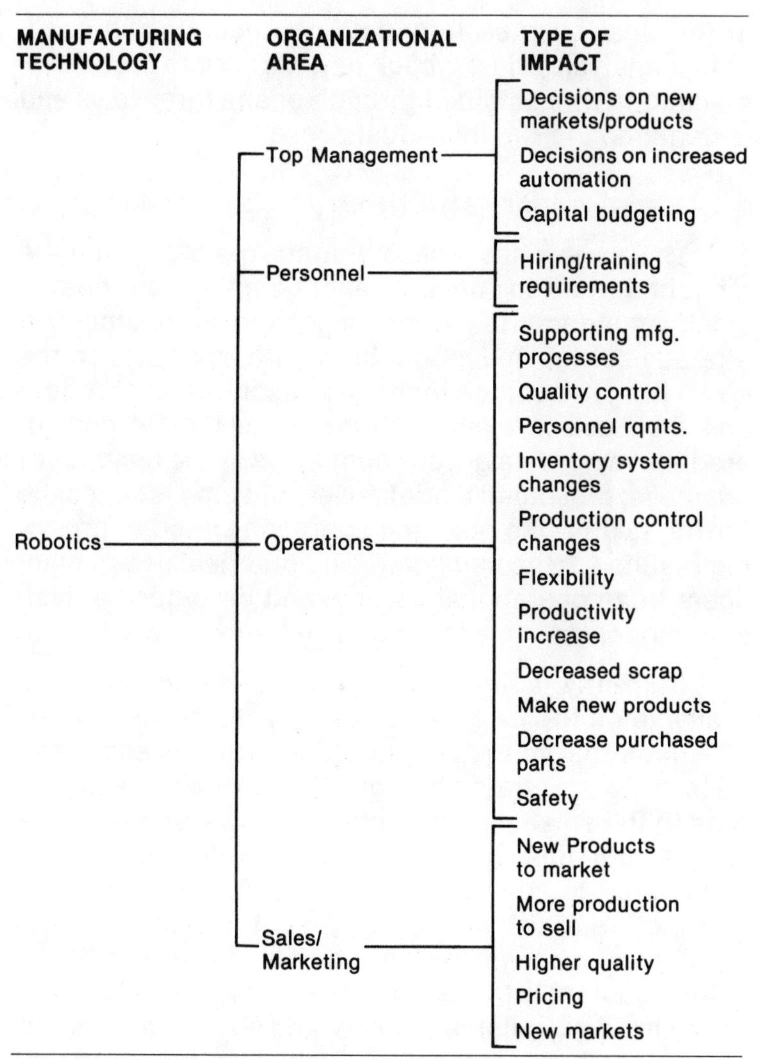

Fig. 2.4. Potential Impacts of Employing Robotics.

used to collect data to identify strategic and tactical factors considered by the firms reported on in this study. The questionnaire did not specifically identify the reasons for adopting robots as strategic or tactical, but responses do allow the drawing of conclusions relevant to the model and a revised model is presented in the final chapter. Information concerning the economic analysis and payback periods for the robotic installations was obtained through onsite interviews and is contained in the individual cases.

Summary

This review has looked at literature believed to be most important to concerns about employing robots in small manufacturing firms. It looked at findings on strategy setting and productivity with emphasis on the small firm; it included technical descriptions of robots and their use in manufacturing processes; it considered the major role government plays in the conduct of business; presented a brief review of some recent calls for the taking of a research approach aimed at providing results for the immediate and practical use of managers in an operational setting; and developed a decision model for the adoption of robotics technology.

Currently, a significant amount of information is available on the technical aspects of robots. However, one finds that in the United States little research has been done on strategic implications of employing robots in the small manufacturing firm. Several possible reasons for this include: (1) the robotics industry is considered to still be in its infancy; (2) producers and distributors feel that despite recent widespread publicity many managers are not aware of the benefits of using robots; (3) the use of robots has been concentrated in a few major industries; and (4) in an attempt to recoup start-up costs robot producers have concentrated on that portion of the market consisting of large firms which can place large orders. Additionally, data collection on the use of robots does not yet differentiate between large and small users. Further refinement of data collection methods on the part of the robotics industry to better determine the distribution and uses of robots is needed as is specific investigation into the use of robots in the small manufacturing firm.

CHAPTER 3

Research Methodology

Introduction

The purpose of this chapter is to describe the research procedure including: (1) the scope of the research; (2) the selection of the research sites; (3) data collection techniques; (4) the case studies developed; and (5) the method of analysis of the data obtained during the research.

The Scope of the Study

This study was undertaken to determine the strategic implications of employing robots in small manufacturing firms. Since a firm's competitive success can be impacted by its use or adoption of new manufacturing technologies [99], the task here was to identify the consequences of employing robots that changed the competitive posture of the business either in a positive or a negative way. The questionnaire used in the case studies identifies internal and external effects of applying robotics. Internal considerations included: labor costs, product quality, output rate, material waste, inventory, and product and/or process flexibility. External effects included the entry into new markets, changing relationships with suppliers, and changing profitability. The research was designed around five questions which were:

1. What are the principal reasons for employing robots?
2. What are the user concerns in employing robots?
3. What are the actual results of employing robots in the small manufacturing firm?

4. What strategic implications can be drawn from the study of the use of robots by the small manufacturer?
5. What are the future plans for employing robots in those firms studied?

The data collected was analyzed for the purpose of identifying common motivators and/or results of employing robots, identifying the strategic implications, and identifying further research needs in the area of the study.

Initial Contacts

In order to help define the scope of the study, interviews were conducted in person and by telephone with individuals active in the fields of robotics development, employment, and education. These experts included two robot distributors, one robotics systems house senior vice president who had over 20 years of experience in robotics, two robotics manufacturing managers, two instructors involved in the training of robotics technicians, and one robot user. These interviews along with a review of the pertinent literature provided the researcher with insights into the following areas:

1. History of the robotics industry
2. Current utilization of robots
3. Status of robotics technology
4. Expected growth rate of robotics
5. Producer problems encountered in transferring robotics to users
6. User problems in employing robotics
7. Perceived advantages of employing robots.

This background confirmed the need for research into the use of robots by small manufacturing firms. One significant finding in both the interviews and the literature, which both limited the scope of this study and reinforced the need for such a study, was the fact that the small manufacturing firms were adopting robotics at a much slower rate than were the large firms.

Research Site Selection

The scope of the research was limited to users of robots who were in the small business category and located in the southeastern United States. A geographic limitation was imposed by limited travel funds. Research site identification was aided by the experts interviewed early in the research. The researcher also received information on possible research sites through correspondence with some 50 robotics producers (both foreign and domestic) with offices in the continental United States, as well as contacts with various universities involved in robotics research projects. This complicated procedure identified only nine potential research sites. The researcher was able to make arrangements to visit seven of these sites. However, only five of the firms agreed to release data for inclusion in the study.

The following comments can be made about the two firms which permitted a site visit, but did not release data for the study. The first firm had one robot which was used for polishing the reflectors of small lamps used in the interior lighting of aircraft. The robot had several times the capacity required for the task, but it was what the "boss" bought according to the engineer who conducted the plant visit. The robot was not in operation during the visit and a man was performing the buffing work by hand. The process consisted of holding each reflector and passing it across the different buffers which were mounted on electric motors. The work caused considerable vibration and was fatiguing which led to lower quality work than had been achieved when the robot performed the polishing tasks. The robot had been out of commission for a week because of the failure of a control system printed circuit board. The equipment manufacturer did not have a spare board in stock and the estimated delivery date of the board was three weeks after the breakdown. The implication of such an occurrence will be discussed in Chapter 6.

The second firm which allowed a site visit, but declined to release data, was engaged in the winding down of tape onto audio cassettes. The operation was highly automated and combined small robots with special purpose equipment and transporters to accomplish the process. The process appeared to be quite efficient and the firm also has an automated warehouse. However, the president of the company decided not to release data on configuration costs, productivity, sales, or profitability.

Aside from the requirement that the selected firms had fewer than 2,000 employees, the following criteria had to be met: (1) cooperation of the owner and/or chief executive officer of the firm was obtained; (2) the firm had robots in operation that were being used in normal production processes, or had been recently employed but were recently discontinued; and (3) the firm had collected or was collecting data which reflected the results of employing robots.

The number of employees in the firms visited ranged from 68 to 1,500. The types of robots employed by these firms were assembly, load/unload, and welding robots. The products of the firms included communications equipment, agricultural mowers, metal doors and windows, hand tools, and transport cases for military weapons. A detailed description of each firm's products is found in the case studies in Chapter 4. The basic nature of the business, its size, and location (state) can be seen in the following chart. Since most of the firms preferred not to be identified by their proper names, they are called Company A, Company B, Company C, Company D, and Company E in this study.

Table 3.1 Firms Included in the Study

Firm	Business	Size	Location	Type Robots*
A	Agricultural Mowers	68	MS	Load/Unload
B	Military Weapons & Support Systems	423	AL	Welding
C	Metal Doors/ Windows	1300	GA	Assembly
D	Telecomm.	1500	VA	Load/Unload & Assembly
E	Hand-tools	450	SC	Load/Unload

*Firms A and C had welding robots in place and in the checkout phase, but no operational data had been obtained.

Data Collection Techniques

This study was conducted through field research and the method was selected because: (1) the small sample size was not conducive to mail survey; (2) the applications of robots differed from location to location even though they were of the same technical type, such as differing use of welding robots based upon the material welded; (3) the use of robots was a relatively new phenomenon in the United States, and historical data on the use of robots by small manufacturers was limited; (4) the desire to produce a practitioner oriented study indicated that the questionnaire developed as a research instrument be supplemented by onsite interviews and firsthand observation; (5) the observation of robots in the working setting provided the researcher with an indication of the degree of confidence that could be placed in the data collected. The researcher had education in electrical, mechanical, and industrial engineering as well as work experience in electronics and metalworking.

The questionnaire for collecting data for the study was designed with the following factors in mind [65]:

1. Decisions about the questions included:

   Are explicit answers desired or will inference satisfy the situation?
   Is the question necessary? How is it useful?
   Does the question cover the group intended?
   Do the respondents have the necessary information to answer the questions?
   Is the question biased?
   Will the respondents give the information that is asked for?
   Can the question be misunderstood?
   Does the question adequately express the alternatives with respect to the point?
   Is the question wording likely to be objectionable to the respondent in any way?

2. Decisions about the form of the question:

   Can the question best be asked in a form calling for a check-answer, free-answer, or check-answer with followup answer?
   If check-answer, what is the appropriate type—dichotomous, multiple-choice (cafeteria) or scale?
   If checklist, does it adequately cover alternatives without overlapping and is it in a defensible order?
   Is it of reasonable length?

3. Decisions about the placing of the question in the questionnaire.

   Are the questions in a logical sequence?

Do general questions precede specific questions?

Since the information sought in the study was explicit, the questions were designed to elicit specific data or were presented in a cafeteria format which allowed the respondent to rank order preferences. Also, the "Not Applicable" category was provided so that a respondent was not forced into assigning an artificial ranking to an item that may not have been considered. There was the requirement to supplement certain explicit data with open-ended questions. For example, although a specific question was posed concerning the financing of equipment purchases, the respondent was also asked to give an opinion of whether financing capital equipment such as robotics is seen as a major problem for small manufacturers.

Efforts were made to ask questions which were necessary to the research and seek data which should normally be available to the CEO and/or staff members. Interviews were conducted with the CEO in some cases and with staff members recommended by the CEO in others. The positions of the persons interviewed are identified in the case studies.

The Questionnaire

The questionnaire (see Appendix A) contains six sections as outlined below:

1. General Information. This section contains questions concerning the firm's industry, identification of person(s) completing the questionnaire, sales, volume, number of employees, production processes, type robots employed and the cost of the robots, and financing of robot purchases.
2. Strategic Factors Considered in Robot Employment Decision. Questions in this section relate to the management decisions which initiated

interest in robotics and the roles of managers in the decision-making process as it related to the use of robots.

3. Principal Reasons for Employing Robots. This section requests a rank ordering of the factors which motivated the use of robots including product quality, labor, output rates, scrappage, and other factors typically associated with the production process. Some of these can be considered to translate into strategic impacts if the results lead to outcomes such as serving of new markets.
4. User Concerns in Adopting Robots. This section deals with objections on the part of management and other employees to the installation of robots and how these concerns were resolved.
5. Results of Employing Robots. This section addresses a wide range of issues and the results of employing robots in the manufacturing processes. Included are identification of the impacts on quality, costs, scrappage, labor and management positions, training and retraining, organizational structure, and worker problems.
6. Future Plans Concerning Robots. This final section of the questionnaire elicits information concerning whether the firm plans on employing more robots in the future.

The Case Studies

Information obtained from interviews and the questionnaire was developed into a brief case study for each firm visited. The cases provide: (1) a company profile or brief history and a description of its products, markets, and available financial data (It should be noted that the small firms are reluctant to reveal finan-

MOTIVATION	Rank Order of Factor*					Average
FACTORS	Co.A	Co.B	Co.C	Co.D	Co.E	
Reduce Labor Cost	1	1	9	1	NA	3.00
Other factors from the questionnaire are entered on successive lines.						

*The rank order system is such that 1 is the most important factor, 2 the second, and so forth.

Figure 3.1. Data Analysis Format.

cial information.); (2) a description of the types of robots being used and the operations being performed by the robots; (3) strategic factors considered; (4) principal reasons for employing robots; (5) user concerns in adopting robots; (6) results of employing robots; (7) future plans concerning robots; and (8) strategic implications.

Data Analysis

The data collected was analyzed to determine the strategic implications of employing robots in the small manufacturing firm. This analysis was done through identification of those results of robot employment which provided potential long-term impacts on the firm's operations such as quality improvements, new markets, changes in labor requirements, and product flexibility. The reasons motivating the adoption of robots were compared across the companies in order to determine whether some factors were consistently considered more important by users than others. Figure 3.1 provides an example of the analysis format.

The implications drawn from the study are then compared with findings from a study of the employment of robots in small Japanese manufacturing firms.

The data analysis is presented in Chapter 5 and forms the basis for the conclusions and recommendations for further research found in Chapter 6.

It is believed that the methodological approach used in this study fulfills the key research needs identified by Thomas and Tymon [125] of descriptive relevance, goal relevance, operational validity, nonobviousness, and timeliness.

CHAPTER 4

The Case Studies

The five case studies presented in this chapter are based upon information obtained during visits to each of the research sites as well as follow-up telephone conversations. The interviews were with the president and/or CEO of the firm or designated representatives. The actual names of the companies, their personnel, or exact location are not revealed at the request of the majority of the participants. The cases are brief because they are oriented specifically toward the employment of robots and because the firms visited are reluctant to release financial data on their operations. However, they were cooperative in furnishing information concerning their reasons for employing robots and data on the results of their use.

The firms described in the cases are in different industries: agricultural, commercial building components, defense, hand tools, and telecommunications. All five are located in the southeastern United States (Mississippi, Alabama, Georgia, Florida, South Carolina, North Carolina, or Virginia). Four of the firms are nonunion and the other considers its relationship with the union such that no problems will occur which cannot be easily resolved by management and union personnel at the site. The number of employees in the plants ranges from 68 to 1,500.

The cases are arranged to provide:

- Company Profile
- Types of Robots and Applications
- Strategic Factors Considered in Employing Robots

- Principal Reasons for Employing Robots
- User Concerns in Employing Robots
- Results of Employing Robots
- Future Plans Concerning Robots
- Strategic Implications

The items identified as reasons for adopting robots form a composite of advantages of robotics typically found in robot manufacturers' advertisements and research studies on robotics. For comparison purposes, the results section follows the format of the reasons given for employing robots. The strategic implications are based upon the results of employing robots such as the entering of new markets, significant increases in quality, output, or other factors which may have a long-term impact on the firm. An analysis of the data from the case studies will be found in Chapter 5.

Case 1
Company A

Company Profile

Company A was founded in 1964 by the current president of the firm. Ninety-five percent of its output consists of agricultural and industrial rotary mowers of the type commonly called "bush hogs." The 65 workers in the plant produce over 200 of these mowers by operating the plant with two shifts per day for five days each week. The mowers are manufactured in five different models and each model comes in six different sizes. The other product is a post hole digger. Both the mowers and post hole diggers are operated via the power takeoff shaft of a tractor.

Company A serves an international market. However, the president, who provided the information for this case, would not divulge the customer mix. He did state that his customers include both farm equipment dealers who market the mowers under the Company A label and other producers of agricultural equipment who market them under their own label. The managers and workers at Company A were proud of the fact that in early 1985 the company won a bid for six months of mower production from a Canadian farm implement dealer. The significance of this contract was that competitors included mower manufacturers from Japan, Brazil, France, Italy, and West Germany as well as U.S. producers. Today, Company A ships mowers and post-hole diggers to 33 foreign countries.

The market share held by the firm is difficult to determine because of the manner in which product data is reported and because a number of the producers, including Company A, do not report sales data to any of the national organizations which collect data on the farm implement industry. The president feels that Company A has a domestic market share of approximately 50 percent. A manufacturers association representative recently informed him that Company A may have an 80 percent share of the market. The company also does not divulge financial information concerning its profitability. The president did indicate that sales for 1983 and 1984 were approximately $10 million and $12 million respectively. This 20 percent increase in sales is significant because the annual growth rate of the industry is only 4 percent.

Types of Robots and Applications

Loaders/Unloaders

In January 1983, the company installed four robots made by GMF Corporation. These robots load and unload CNC lathes which turn ductile iron into universal joints, bearing caps, shafts, blade bolts, hubs, and all gear box components for the rotary mowers and post hole diggers. Total installed cost of the four robots was $110,000.

The robots are configured in work cells which include the robot, the CNC machine which the robot tends, and a feeder pallet which feeds stock to the robot and receives the finished product unloaded by the robot, plus a human worker who serves as quality control monitor, transfers the finished product to transport baskets, and fills the pallet with stock. Each robot has two grippers on the end of its arm. Each gripper can handle a ten pound load. By the time the lathe cycles end, each robot is positioned to remove a finished part from the chuck and to insert a blank. Within seconds, the robot arm removes the finished part, the hand rotates and inserts a blank, the arm retracts, and the lathe recycles. While the lathe recycles, the robot places the finished part on an empty pallet, and picks up a new blank to place in the lathe on the next cycle. Feeder pallets, 20 per station, serve parts to three of the robots. A 30 pallet feeder supports the other robot. Each pallet is equipped with a Z-axis motion to raise parts to the robot gripper or hand. The two gripper configuration of the robot hand saves time in loading and unloading the lathes by allowing the robot to deposit a blank into the lathe chuck before retracting to deposit the finished part on the pallet.

The feeder pallet systems in the work cells enable the robot supervisor to check the finished parts for quality before loading them into transport baskets. When the parts index around the table on the pallet, it is easy to spot a flaw. If a flaw is found, the operator can stop the system and change the cutting tool to minimize scrap parts.

The product mix mentioned earlier causes a considerable variance in the production needs. This means that each lathe and robot must accommodate two or three parts changeovers each week. According to the company president,

> Within 15 minutes, one of our technicians changes a program for a robot and its grippers, adjusts the lathe chuck, and is ready to process another set of parts. This is simple compared to the old automatic lathes! It used to take a whole day to set up for another part, then another day to get the part running right. Now we can finish nearly an entire batch in that time.

Company A added versatility to the work cells by having the robots mounted on unique adjustable brackets. These devices reposition the robots in any direction—up, down, forward, backward, or side-to-side—to accommodate wider or longer parts.

Welding Robots

On April 11, 1985, two Shen Meiwa seam tracking welding robots were placed into operation. These robots weld the rotary mower frames. The welding robots are set on each side of a work table and weld the frames of the mowers after they are manually aligned and placed into position. While one frame is being welded, the next is aligned, and then slid onto the welding table as the finished frame is removed. The weld positions are taught by a technician who programs the weld by moving the welding head through its welding positions while entering them in the controller program via a hand held programmer. It takes an experienced programmer about 30 minutes to reprogram the robots when a different size mower body is to be welded. The wrist motion of the robots allows them to weld seams of varying length on vertical or horizontal surfaces as they perform their programmed tasks. At the completion of the weld, the head moves to an automatic cleaner which cleans the tip prior to the next weld. The welding robots were in the checkout phase when this case data was collected and the first weld time was 7 minutes and 40 seconds. When working at designed operating speed, the robots will weld a frame in 3 to 4 minutes. The manual welding operation requires 24 man-minutes (two welders working 12 minutes each). The welding robots and associated equipment cost $250,000 installed and checked out and including training for Company A programmers and loaders. More information on this and the costs associated with the other robots will be presented in the section on the results of employing robots.

Strategic Factors Considered

The competitive strategy of Company A is to produce the highest quality product in the industry at a competitive price. When the CNC lathes and a CNC milling machine were placed in the plant in the mid-1970s, they had a significant impact on productivity. This favorable impact convinced management that further automation could create greater competitive advantage. In 1982, the president of the firm saw robots on exhibit at a machine tool show in Chicago and felt that the robots could help improve productivity in his plant.

Principal Reasons for Employing Robots

Beginning with the most important factor, the following were ranked in order of importance to Company A as reasons to employ robots:

1. Reduce labor costs
2. Increase output rate
3. Relieve a shortage of skilled workers
4. Relieve workers of tedious and/or dangerous jobs
5. Reduce capital costs (more CNC machines and attendant labor would cost more than the robots)
6. Improve product quality
7. Remedy high labor turnover problems

8. Decrease material waste
9. Increase flexibility in changing products, designs, etc.
10. Ease compliance with OSHA regulations

User Concerns in Adopting Robots

It was reported that Company A experienced no problems or complaints from the workers when the plans to introduce robots were announced. The employees were informed that some would be trained as operators and maintenance personnel for the robots.

Results of Employing Robots

Company A found it easy to identify positive results from initiating robot use. At the outset of identifying results, it should be pointed out that no worker lost a job because of automation. In fact, the work force increased by 20 people. Now, however, 5 workers and the four loader/unloader robots perform the operations and quality control inspections that would have required 45 workers with the system as it was before robots were installed. The following quote is from the owner:

> The work is much easier too. Supervising the robots is actually the most prestigious job in the shop. When we put the robots on the lathes, we found we produced 50 percent more parts per eight hour shift. That increase was most welcome. In the past few years we had to buy half our parts to keep up with the rest of the plant. Now even some of our competitors buy our components because it costs us so little to manufacture a surplus.

Calculations on scrappage indicate that less than one-half of one percent of the parts produced with the robotized operations end up in the scrap pile. In the past, scrap rates had hovered around 10 percent. The above information along with additional data is recapped in Table 1.

The welding robots cost a total of $250,000 for the pair as stated above. The cost of welding labor per mower under the manual system is $3 for an annual cost of $144,000 with normal production. At this rate and excluding maintenance costs, the payback period for the welding robots would be:

$$\$250{,}000 \div \$144{,}000 = 1.736 \text{ years.}$$

So it is estimated that the welding robots should pay for themselves in about two years if they perform as expected.

The percentage of operational time for the welding robots is not known, but is estimated by the manufacturer to be around 98 percent. The loading/unloading robots have achieved over a 98 percent rating for operational time. The systems costs for these robots were:

One robot system at $35,000 (robot—$16,000 + $19,000 associated equipment)

Three robot systems at $25,000 each (robot—$15,000 + $10,000 associated equipment)

The total cost for the four systems was $110,000. The payback period for the load/unload robots is shown in Table 1. The president of Company A credits the robots with enabling him to compete in the global market on a successful basis. The degree of that success may be indicated by the fact that Company A paid cash for all of its robots.

Table 4.1 Results of Employing Robots at Company A

Productivity—50% increase in component production per shift.
—20% increase in complete unit production per shift.

Scrappage—Twenty-fold decrease in scrap, from 10% to 0.5%.

Table 4.1, continued

Purchased Parts—Produces all needed parts in-house plus sells components to competitors. Previously bought 50 percent of components from outside sources.

Labor—25 new positions created; 5 filled by retraining incumbents, 20 new hires.
- —Labor costs reduced by 30 percent per unit of finished product.
- —Relieved shortage of skilled labor.
- —Relieved workers of tedious and/or dangerous jobs.
- —Reduced work-related accident rate.
- —Operative skills have increased.
- —Maintenance skill levels have increased.

Payback period on the GMF robots—Company A estimate is 75 days.

Minimum payback period considering only labor savings of 20 workers not required by new system compared to old.

$$\text{Payback} = \text{Robot Costs} \div \text{Labor Savings}$$

(20 workers) ($5/hr) (8 hr/shift) (2 shifts) = Labor savings*

Labor savings = $1600 per day.

Payback = $110,000/$1600 = 68.75 days

*Savings of 20 workers based on 5 workers doing job now versus 45 under old system = 40 workers less 20 new hires = savings of 20 workers.

Unexpected Consequences:

1. Company A no longer had to depend upon outside supplies for finished parts.
2. The company became a supplier of parts to its competitors.

Future Plans Concerning Robots

The president of Company A is a strong believer in automation. To him it is conceivable that he and his three sons could at some future date place the entire plant in operation on a given morning just by pushing a few switches. The policy of accumulating cash for all equipment purchases may hinder the speed with which more robots or other automation are installed.

Strategic Implications

The strategic implications of employing robots at Company A are significant. The following things have occurred:

1. Both output rate and quality are up significantly.
2. A sales growth of 20% was obtained in a market with overall growth of only 4%.
3. The company is no longer dependent upon outside suppliers for finished parts.
4. The company has entered new markets as a supplier of parts to other manufacturers in the industry.
5. Company A competes successfully in the international market.
6. This case illustrates that there are potential applications for robotics which have an extremely short payback period.

Case 2
Company B

Company Profile

Company B was founded in 1963 and describes itself as a 100 percent provider of military equipment and the best subcontractor to the U.S. aerospace industry. Products manufactured by the firm include shipping and storage containers for munitions such as air-to-air missiles, cluster bombs, and land mines; cluster bombs, stabilizer fins for missiles and bombs; and a helicopter transport trailer. The company's product services range from design and engineering to product delivery as well as after sale service. The company has 423 employees.

Types of Robots and Applications

Company B employs two arc welding robots. One is a Cincinnati Milacron robot which has five axes of movement and can weld a continuous seam up to 50 inches long. it is used to weld both inside and outside seams on an aluminum shipping container which measures approximately 48 by 24 inches. This robot was installed in January 1983 at a cost of $118,000. The other robot was installed in January of 1984. It is smaller than the Cincinnati Milacron robot and is used for welding smaller components made of steel. This robot was manufactured by ASEA and had an installed cost of $90,000. It is interesting to note that the purchase of both robots was through the use of municipal industrial bonds.

Strategic Factors Considered

Interest in robotics was initiated because of a desire on the part of senior management to reduce labor costs in the firm which is a typically labor intensive metals fabrication operation. The specific personnel involved in making the decision to install robots were the president, vice president for finance, the production superintendent, and the manufacturing manager. Their roles were to identify needs and justifications for the use of robots. Aside from labor costs, product quality improvement was considered as a factor in maintaining a competitive position in the defense industry. The two robots are now featured prominently in company advertising brochures which tout the advanced manufacturing technology employed by the firm. Company B feels this is a justifiable marketing strategy since it is one of the few small firms, particularly in the Southeast, using robots.

The decision to proceed with purchase of the first robot was based upon data provided by the robot producer and visits to a large manufacturer who reinforced the claims of the robot manufacturer. The expected savings in labor amounted to 20 to 30 percent. Other decision factors are presented below.

Principal Reasons for Employing Robots

Company B considered the following factors to be the most significant potential benefits to be gained from employing robots. The factors are ranked from most to least important.

1. Reduce labor costs
2. Improve product quality
3. Increase output rate
4. Relieve workers of tedious and/or dangerous jobs
5. Decrease material waste
6. Increase flexibility in changing products, designs, etc.
7. Reduce capital costs
8. Remedy high labor turnover problems
9. Relieve shortage of skilled workers
10. Ease compliance with OSHA regulations

User Concerns in Employing Robots

Management informed personnel that robots would be installed in the plant about two months before expected delivery of the first robot. The employees are unionized and union leaders expressed concern over a possible loss of jobs because of the robots (management considers the union in the plant to be a relatively weak organization). These problems were resolved through assurance by management that no jobs would be lost and that employees would be trained to operate and maintain the robots. However, management was concerned about possible sabotage of the robot and under the guise of a safey measure, a fence was installed around the robot work cell and access was limited. Within six months, it was apparent that there was no opposition to the robot and the fence was removed. Management had no such misgivings about introducing the second robot.

Results of Employing Robots

Table 1 provides a summary of results of employing robots at Company B. Aside from the significant savings in labor and the increased output rate, a strategically important result was the response of senior management. The president of Company B directed the production superintendent and manufacturing manager to proceed with investigating the possibility of further automation including the feasibility of installing a CAD-CAM system. Additionally, management agreed to proceed with budget planning for the installation of from two to four more robots during the next year. Both welding and painting robots are being investigated for the next installations. This "new management thinking" was brought about by the impressive cost savings and productivity increases generated by the two welding robots.

Future Plans Concerning Robots

As mentioned above, the positive results achieved thus far have caused senior management at Company B to become aggressive in pursuing automated manufacturing processes including installation of more robots in the immediate future and investigation of the feasibility of installing a CAD-CAM system.

Table 4.2 Results of Employing Robots at Company B

Productivity—500 percent increase in output rate

Scrappage—1 percent decrease in scrap.

Purchased parts—No change, parts were already made in house.

Labor—20 new positions created; 18 filled through retraining of incumbents, 2 by new hires.
—Labor costs reduced by 40 percent per unit.
—Relieved workers of tedious/dangerous jobs.
—Relieved shortage of skilled labor.
—Operator skills have increased.
—Maintenance skills have increased.

Payback period for the robots—Estimated payback period was one year for each robot. Exact data was not available.*

*Management felt a shorter payback would have been possible if they had anticipated the significant increase in output possible with the robots. Some 90 days were consumed in bringing support facilities up to the speed of the robots.

Unexpected Consequences:

1. Senior management reaction—the increased productivity credited to the installed robots sensated what the Firm calls "new management thinking." That is a dedication to support of increasing plant automation, including the establishment of a Robotics Department to plan automation projects and an operating budget to support planning for CAD-CAM system.

2. Need to increase support capacity in other parts of the manufacturing process because of the high productivity of the robots.

Strategic Implications of Employing Robots

Company B's experience with robots has caused management to take a serious and long-term look at moving toward a flexible manufacturing system for its plant. The combination of higher quality and lower manufacturing costs coupled with additional flexibility has caused management to think in terms of a total systems concept, from design to finished product, about its operations. Although the firm sees an increasing ability to compete in several markets, senior management believes strongly in its niche strategy and does not intend to pursue commercial markets at this time.

To emphasize its seriousness about proceeding with automation, the firm has altered its structure by establishing a Robotics Department. Presently staffed by a manufacturing supervisor and an industrial engineer, this department will spearhead the company's investigation of flexible manufacturing alternatives.

Case 3
Company C

Company Profile

Company C was founded in 1960 and the founder is the current president and chairman of the board. The firm has 1,500 employees in three plants. Sales for 1983 and 1984 were $100 million and $122 million respectively. Company products are steel framed windows and doors used primarily in commercial buildings. Doors are produced in a plant located in the Southeast at the same site as the company headquarters. A plant 25 miles away produces windows. A larger plant is located in the Midwest; both doors and windows are manufactured there. Products are produced to order with a typical annual production breakdown as follows:

10% custom (1–100 units per year)

10% small batch (101–1,000 units per year)

70% large batch (1,001–10,000 units per year)

10% mass (over 10,000 units per year).

Company C is an industry which has approximately 300 competitors and estimates its market share to be between 15 and 20 percent. Because entry barriers are typically low and firms enter and exit the market with economic cycles, the reporting of data on the competition is not precise. Raw materials, sheet steel and glass and other components are readily available and any firm with a sheet metal fabrication ability is a potential competitor.

At this firm, the president referred the researcher to the general manager of the door plant and the senior analyst advanced technologies for the interviews and collection of data. The latter is responsible for automation projects at both southeastern plants.

Types of Robots and Applications

Company C's midwest plant has no robots installed, so this discussion centers around the two southeastern plants. The door plant recently installed two Devilbliss plasma arc welders for welding seams of doors. This equipment was still in the checkout phase and operational data was not available. Systems costs for the welding robots was $140,000.

The robot at the window plant was installed in January of 1985 and is considered to be an assembly robot. Company C calls it a hybrid robot because it contains parts from different robot manufacturers and the system was designed by a robotics systems house. The task of the robot is to apply a silicon glazing compound to one side of a window, turn it over and apply the glazing to the other side, and then place it on a pallet for transport to the next station. The installed cost of this system was $275,000. Company C financed the purchase of its robots through a third party lease arrangement.

Strategic Factors Considered

The president and executive vice president of the company were credited with initiating the interest in robots. The senior analyst advanced technology was charged with the responsibility for investigating robotics applications. The major strategic concerns of Company C which brought about the interest in robotics revolved around three competitive factors. The first was that the economy was in a recovery mode with a great deal of new construction taking place in the major metropolitan area the company served, as well as in the Southeast in general. The president was concerned about a potential loss of customers if output could not be increased to meet demand. The alternative of using overtime was not considered attractive because of the cost and the fact that workers were already on a ten-hour work day. Both of the southeastern plants are on a four-day work week. The workdays are Monday through Thursday. The second factor was product differentiation through the production of a superior quality product in a very competitive market. The third factor was the potential ability to gain more flexibility in changing product configuration since the firm produces to order.

The concern for quality led to the introduction of the robots at both plants. At the window plant, it was desirable to use a new silicon glazing compound that promised superior finish and durability compared with that which had been used for some time. The difficulty of using the new compound was that its properties were such that a human could not apply a consistent bead. This led to the consideration of robots for the job. The robot has performed consistently and satisfactorily for over six months.

Principal Reasons for Employing Robots

As indicated above, competitive position was a driving factor in the interest of robotics at Company C. The senior analyst advanced technology ranks the following factors from most to least important in the decision to employ robots.

1. Increase output rate
2. Improve product quality
3. Increase flexibility in changing products
4. Decrease material waste
5. Relieve workers of tedious/dangerous jobs
6. Remedy high labor turnover problems
7. Relieve shortage of skilled workers
8. Reduce capital costs
9. Reduce labor costs
10. Ease compliance with OSHA regulations.

The quality and output rate concerns in both the door and window plants were related to the fit and finish of the products. At the door plant the present assembly method uses three to four spot welds at each of four vertical seams. Two are at the top and two at the bottom of the door. After the doors are assembled, they are filled with a liquid foam which expands as it hardens. Some leakage occurs between the spot welds which leaves a foam deposit on the outside of the door along the seams. This is scraped off with a razor blade and/or a scalpel like knife. Occasionally, the cleaning process results in a scratch in the prime coat on the door frame. The company ships the doors primed and the customer applies the finish coat. A scratch in the prime coat could be a source of rust, hence possible rejection by the customer. The seam welder would provide a seal at the door seams which would prevent leakage of the foam filler and the cleaning process should be eliminated. The four to six workers at the inspection and cleaning stations could just perform inspections if the seam welder proves successful. This procedure should eliminate a bottleneck in the production line. The robot welding system would be approximately 30 percent faster than the spot welder.

At the window plant, the primary concern was the inability of humans to apply the new glazing compound in a consistent bead. Here, the robot is consistent and less cleaning and rework is required.

User Concerns in Employing Robots

Company management did not provide advance notice to the workforce that robots would be placed in the plant. When the robots were introduced, it was explained to the workers the reason was to enable the firm to remain competitive in its industry. Company C management stated that the workers voiced no objections to the addition of robots, nor was there negative feedback from supervisors. Company personnel were informed that some of them would be trained to operate and maintain the robots. As it turned out, three new positions were created at the assembly plant as the result of employing the robot there.

Results of Employing Robots

As mentioned above, the welding robots at Company C were in a checkout period and operational data was not available. The assembly robot was designed to perform a task which could not be performed consistently by a human. The company has not measured the output rate of the robot system versus that of humans using the old glazing compound. However, the company is satisfied that the robot is performing its designed task as it should be.

Approximately 120 days after the original data was collected, management at Company C decided that the plasma arc welding process could not be used for its product and the welding robots were removed. The reasons given for failure of the process were that the welding technique generated too much heat for the sheet metal and caused buckling. The firm then turned to the investigation of alternative welding methods such as laser welding. Increased use of robotics and/or other automation remain a high priority interest in the firm.

No workers lost jobs at Company C when the robots were installed. New positions were filled by re-

Table 4.3 Results of Employing Robots at Company C

Productivity—The robot enabled the assembly plant to windows of superior quality compared to those produced by humans.

Scrappage—Not directly measureable; see comments under "Productivity" above.

Purchased Parts—No change.

Labor—Two new management positions, four new hourly worker positions were created. All positions filled by encumbent employees.

Payback Period for the Robots—Not calculated, the need for the technology in the assembly process was the justification.

Unexpected Consequences—None for the assembly robot. Welding robots were removed from the plant because of unsatisfactory performance. It should be pointed out that the welding method and not the robots or control system caused the removal.

training workers already in the plants. These positions are two lead operators, one computer programmer, and one welder. Two new management positions were created with the interest in robotics. They are discussed in the strategic implications section.

Future Plans Concerning Robots

As discussed in the next section, Company C plans to pursue further employment of automation. However, at the time data was collected for this case, no specific robot applications had been identified.

Strategic Implications

The employment of robots by Company C was based on the strategic concern for maintaining and increasing market share through a strategy of high quality and on time delivery. A significant implication for the future direction of the company can be found in the establishment of the position of senior analyst advanced technology and a new manufacturing engineering position. The senior analyst is responsible for implementing a management information system as well as investigating, with the assistance of the manufacturing engineer, the utilization of more advanced manufacturing technologies in the southeastern plants. The president of Company C wants to be aggressive in the approach to quality and on time delivery. The means of improving performance in these areas is seen as the employment of advanced manufacturing technologies. These efforts could also produce entry barriers to potential competitors who could not afford to automate their processes.

Case 4
Company D

Company Profile

Company D is in the telecommunications industry and produces telephones which range from basic desk sets to those with multistation interconnect capability. The company employs 1,300 people with 1,000 located at the headquarters plant and 300 at a plant some 50 miles away. Since the company molds its own plastic telephone cases and components and manufactures printed circuit boards along with other parts for the telephones; it runs both continuous process and assembly line operations. Its customers are wholesalers and retailers who sell telephones and telecommunications to companies which install complete commercial systems. Fifty-three percent of production is made-to-order and the remainder is made-to-stock.

At present Company D is operating at approximately 60 percent of capacity because the company had planned on marketing the bulk of its output through AT&T telephone centers. However, the AT&T divestiture and the influx of telephones sold by independent producers greatly reduced the projected sales of Company D. In its attempts to deal with department and discount stores, Company D has had difficulty competing with lower cost imports. The sales revenues for the firm were approximately $90 million and $95 million in 1983 and 1984 respectively. At present, the company is actively pursuing more business from those firms which sell complete commercial communications systems and see this as a significant portion of future business. Company D recently bought a small (100 people) firm on the West Coast which makes circuit boards, but no production has been received from the plant at the time data on Company D was collected.

The information for this case was provided by the vice president-strategic programs at Company D and a plant tour was conducted by a manufacturing engineer.

Types of Robots and Applications

Company D has six robots which are utilized as follows: Three robots unload injection molding machines which make plastic ear and mouth pieces, hand sets, and cases for telephone instruments. The robots place the parts on conveyor belts for movement to inspection and transport stations. The cost of each robot installation was $75,000. Two were installed in June of 1983 and the third in January 1984. One assembly robot attaches the plastic feet to the metal base of the telephone instrument. This robot had an installed cost of $50,000 and was placed in operation in January 1985. All four robots described so far are located in the main plant.

The smaller plant has two assembly robots which work as a team at one work station. Their task is to assemble a spring assembly with a printed circuit board attached. One robot twists the three spring detents to the required 45 degree angle to secure them to the circuit board, then the second robot applies a solder mask material from a dispensing needle attached to its arm. These robots were installed at a cost of $50,000 each and are expected to have a one-year payback period. All of the robots at Company D were purchased through bank financing.

Strategic Factors Considered

The impact of foreign competition was the major cause of generating an interest in robotics at Company D. The manufacturing engineers suggested the use of robots as a method of lowering production costs. The president, the chief executive officer, and the director of advanced engineering were the main actors in the decision to employ robots. The director of advanced engineering and his staff investigated application possibilities and costs of alternative systems, the president acted as review authority in comparing the return on investment for robotics versus other projects initiated in the company, and the chief executive officer was responsible for arranging financing.

For Company D, the most critical factor was seen as U.S. labor costs versus labor costs of producers in the Far East. Reducing labor costs while improving quality through the use of robotics appeared to be a viable strategy for remaining in a competitive position as well as expanding its business with those firms which provide turnkey communications systems.

Principal Reasons for Employing Robots

The major reasons for employing robots at Company D were rank ordered as follows:

1. Reduce labor costs
2. Improve product quality

3. Increase output rate
4. Decrease material waste
5. Increase flexibility in changing products
6. Ease compliance with OSHA regulations
7. Relieve workers of tedious/dangerous jobs
8. Relieve shortage of skilled labor
9. Remedy high labor turnover problems
10. Reduce capital costs.

User Concerns in Employing Robots

The employees of Company D were informed of the pending robot implementation about 90 days prior to the scheduled installation. There were no objections from managers, but some of the direct labor personnel believed that the robots would eliminate their jobs. The managers and foremen of this nonunion plant held meetings with the work crews to explain the purpose of installing robotics was to enable the company to be competitive and thus to actually help ensure continued employment for the workers. Employees were informed that no one would lose a job because of the robots, but that some would be retrained for new positions within the plant with some of those positions being in the robotics maintenance and operations areas. These actions overcame the concerns of the direct labor personnel about their job security.

Results of Employing Robots

As indicated in the applications section, the robots at Company D are performing tasks related to producing parts which are assembled by human workers on an assembly line. The primary justification insofar as the company officials were concerned was labor savings. The plants operate three shifts, five days a week, fifty weeks a year. The entire facility closes one week in the summer and one week at Christmas.

Each of the unloading robots reduced labor requirements from two persons per shift for unloading to one person per shift. For all three shifts, this meant that labor savings were the cost of three workers per day. The calculations in Table 1 indicate a 2.5 year payback period for the unloaders considering the labor savings alone. The firm has not assigned a monetary value to increased quality (yield per shift), decreased material waste (5 percent), or quality control inspections now being performed by operators.

The assembly robots are expected to produce labor savings including a one year payback as well as a saving of $2,000 in solder mask material each year. Also, the assembly robots are expected to provide a degree of flexibility, but the unloading robots are seen as less flexible than humans. However, both types of robots have been dedicated to specific tasks thus far.

The molding unloaders have definitely eased compliance with OSHA regulations. When human workers unloaded the molded parts they were required to wear safety equipment such as gloves and masks. These are not required of the operator when the robot does the unloading. Table 4.4 summarizes the results of employing robots.

Future Plans Concerning Robots

Company D management feels that more robots will be installed in the future, but specific applications have not been identified. Studies are being conducted to identify assembly operations which are candidates for automation.

Strategic Implications

The employment of robots at Company D resulted from competitive pressures. This response to the competition has placed the firm in the position to be active in seeking new markets based upon the company's ability to integrate advanced manufacturing technology into its operations. Company D recognizes the

Table 4.4 Results of Employing Robots at Company D

Productivity—50% increase in parts production based on labor savings.

Scrappage—5% reduction in scrappage

Purchased parts—Not applicable

Labor—The six robots reduced labor requirements by 18 persons. No one lost a job, but people were retrained for robot related and other jobs.
—Ease compliance with OSHA regulations
—Relieved workers of tedious/dangerous jobs.

Payback Period—Calculated for one unloading robot, based on labor savings only.

Payback = Robot cost ÷ Labor savings per year
= $75,000/3(50)(200)*

Payback = 2.5 years

Payback Period—Assembly robots: one year plus save $2,000 in material costs.

Unexpected Consequences—None.

*Labor savings = 3 workers per week for 50 weeks at $200 per worker per week.

competitive advantage to be gained through improved productivity made possible by automation.

Case 5
Company E

Company Profile

Company E was founded in 1895 in the state of New York. The operations were moved to a rural southeastern location in 1962. The company manufactures specialized pliers for industry, commercial pliers, adjustable wrenches, torque wrenches, torque screwdrivers, testers, and calibrators. The product line consists of 200 standard plier models, 1,000 custom models, and over 1,200 different automotive hand tools. The products are sold under two company brand names and several private brand names.

Company E is operated as a wholly owned subsidiary of an automotive tool and equipment firm headquartered in Connecticut. Total employment of the firm is around 1,500 with 450 of these being located at the Company E plant. Company E occupies a 400,000 square foot plant which is fully air-conditioned. Equipment to manufacture the line of hand tools includes forge hammers, broaches, electrojet die sinking machines, milling machines, polishing machines, electroplating equipment, and edge filers. The edge filing is a manual operation in which plier cutting edges are filled and aligned prior to hardening. The production lot quantities of the company are typically in the large batch (1,000 to 10,000 units) category.

Types of Robots and Applications

In February of 1985, Company E installed three loader/unloader robots at a total cost of $250,000. The job of the robots is to place the hot plier pieces in trim presses after they are unloaded from the forge presses. This process requires picking up the forging and placing it in a trim press with proper orientation, depending upon whether it is a right hand or left hand side of the plier, and then unloading the piece from the trim press after it is trimmed. Magnetic end devices were selected by the robotics producer as most appropriate for this particular operation. However, after eight months the robots have not proven to be capable of operating on a consistently successful basis. The primary problem seems to be in the magnetic end device which has difficulty in placing the plier pieces in the proper orientation for the trimming operation. There are several possible technical factors involved such as the effect of the heat of the metal on magnetic properties, strength of the magnets, etc.

At the time of this case the robot arms had been returned to the laboratory of the robot manufacturer for further studies. It should be pointed out that Company E has not abandoned the robots because of a significant strategic factor described below.

Strategic Factors Considered

Company E produces some 1,000 types of custom pliers. These are made to customer specifications and sold primarily to the electronics assembly industry. The manager of engineering during an assessment of future product needs observed that the trend toward more automation in the electronics industry would make some of the hand tools Company E produced obsolete in 10 to 15 years. This observation led to an investigation of methods of adding flexibility to the production line which could aide the transition to other products, but provide current benefits as well. The manager of engineering, a project engineer, and the quality control supervisor worked as a team to identify the appropriate technology which might meet the dual role described above. Robotics was determined to be a technology which could be integrated into the present production line plus help meet the strategic goal of adding flexibility when new products replaced obsolete ones. In Company E one finds a clear example of a perceived strategic need driving the adoption of robotics.

Principal Reasons for Employing Robots

Aside from the long-term consideration described above, the management of Company E identified six factors as being important in the decision to employ robotics. In rank order from most to least important these are:

1. Relieve workers of tedious and/or dangerous jobs
2. Improve product quality
3. Increase flexibility in changing products, designs, etc.
4. Decrease material waste
5. Ease compliance with OSHA regulations
6. Relieve shortage of skilled workers.

User Concerns in Adopting Robots

Company E reported no objections from employees to the robots in the plant because the robots were to perform the hazardous task of handling hot forgings and loading and unloading trim presses, as well as create an opportunity for present employees to be trained as robot maintainers and operators.

Results of Employing Robots

Although the robots are undergoing modification by the producer, Company E felt that advantages had been achieved by the employment of the robots. These were rank ordered as:

1. Increased flexibility in changing products, designs, etc.
2. Relieved workers of tedious and/or dangerous jobs
3. Eased compliance with OSHA regulations
4. Relieved shortage of skilled labor
5. Reduced labor turnover
6. Improved product quality.

In comparing the rank order of reasons and the results of employing robots, it is seen that the factors ranked under results differ from the rationale used for adopting robots. "Decrease material waste" is replaced by "reduced labor turnover." Use of robots has not decreased material waste for the company because of the problems with the robot end devices, but

Table 4.5 Results of Employing Robots at Company E

Productivity—Because of the problems with the robot grippers, production has not increased. However, specifications tightened in anticipation of using robots have been maintained and product quality has improved.

Scrappage—No change.

Purchased Parts—No change.

Labor—Ten new positions related to robotics, filled by training encumbents.

Payback Period—Not calculated, data not available.

Unexpected Consequences—Inability to successfully employ the robots has caused some concern, but the firm intends to stay with the project because of long term considerations.

other positive results have been noted. The manager of engineering commented that the tightening specifications on the product was a major reason for considering robotics and that this has been achieved to some degree.

No changes in the number of employees in the plant occurred, but five individuals were trained as maintainers and five as operators with an accompanying increase in their skill levels. These personnel were trained by the robot producer.

Payback period has not been estimated because of the erratic performance of the robots and the fact that the investment was considered one necessary for placing the firm in a long-term competitive position.

Future Plans Concerning Robots

At the present time, Company E is not committing itself to future employment of robots. It is anticipated that this will be investigated after correcting problems with the present robots.

Strategic Implications

The ultimate goal of Company E insofar as robotics is concerned is to be able to provide a degree of flexibility to meet changing production needs as current products become obsolete. This is a long-term aspect of the use of robots. The results already identified indicate that robots can be a major strategic factor in future competition and the management is open to consideration of future automation efforts.

CHAPTER 5

Analysis of Data

In this chapter the data collected at the field research sites will be analyzed for the purpose of providing a basis for drawing conclusions and making recommendations in Chapter 6. The analysis will address the five research questions presented in Chapter 1. These questions are:

1. What are the principal reasons for employing robots?
2. What are the user concerns in employing robots?
3. What are the results of employing robots in the small manufacturing firm?
4. What are the strategic implications of using robots in the small manufacturing firm?
5. What are the future plans for employing robots in those firms studied?

Comparisons will be made in tabular format of how the different firms ranked the factors which motivated the employment of robots (Tables 5.1 and 5.2) as well as a comparison of the results obtained versus the expected results for the five firms studied (Tables 5.2 through 5.7). In this study the expected results are defined as those factors which were identified as the reasons for employing robots.

The findings in this study of the reasons for employing robots and the results of employing robots will then be compared with the findings of a similar study done in Japan [59].

Table 5.1 Analysis of Reasons for Employing Robots

MOTIVATION FACTORS	Rank Order of Factors					Average Rank
	Co.A	Co.B	Co.C	Co.D	Co.E	
Reduce Labor Costs	1	1	9	1	NA	3.00
Ease OSHA Compliance	10	10	10	6	5	8.20
Reduce Cap. Costs	5	7	8	10	NA	7.50
Improve Quality	6	2	2	2	2	2.80
Increase Output	2	3	1	3	NA	2.25
Increase Flexibility	9	6	3	5	3	5.20
Decrease Waste	8	5	4	4	4	5.00
Relieve Tedious Jobs	4	4	5	7	1	4.25
Remedy Turnover	7	8	6	9	NA	7.50
Relieve Worker Shortage	3	9	7	8	6	6.60

NOTES: 1. The factors are rank ordered with number 1 representing the most important factor.

2. "NA" indicates a factor was not considered. This applies only in the case of Company E. For the factors where an "NA" appears, the average ranking was found by averaging the four assigned rankings.

Other pertinent data from the individual cases will also be examined in this chapter. This data provides information about the impact of robots on the workforce of the firms, methods of financing the robot systems, and the approach to adopting the robots by the firms studied.

The ranking of reasons for adopting robots by the firms visited also provides an identification of their primary strategy. For example, it is seen in Table 5.1 that cost reduction was the primary strategy of Companies A, B, and D. Company C's focus was on the increasing of output while Company E identified relieving workers of tedious and dangerous jobs as a primary objective. Although it was not revealed in the questionnaire, management at Company E stated that overcoming eventual product obsolescence and improving product quality were the main reasons for adopting robotics.

Aligning the factors motivating the adoption of robots based upon the average values, the following is found:

Table 5.2 Reasons for Employing Robots Ranked by Mean Values

Factor	Average Rank
1. Increase output	2.25
2. Improve quality	2.80
3. Reduce labor costs	3.00
4. Relieve dangerous/tedious jobs	4.25
5. Decrease waste	5.00
6. Increase flexibility	5.20
7. Relieve worker shortage	6.60
8. Remedy turnover	7.50
9. Reduce capital costs	7.50
10. Ease OSHA compliance	8.20

Some comments on the ranking of the averages are in order, especially with a small sample. First, in the average rankings, "reduce labor costs" ended up in the third place of relative importance. However, three out of four firms ranked this as the most important factor. Company C which ranked the reduction of labor costs as next to last in importance was concerned with a quality problem which required rework and created a bottleneck in the production line. Solving the quality problem could eliminate two or three people in the rework stage of the assembly line, resulting in decreased labor costs. On the other hand, "Increase output" was first in the average rankings with only one firm choosing it as the most important factor under consideration. These events indicate the extreme impact one value can have on the average of assigned values in a small sample.

A review of the individual rankings by company reveals that some companies had situations not existing or not so severe as other companies. For example, Companies A, B, and C placed the compliance with OSHA regulations as the least important, while Companies D, and E placed them in the middle of important factors. Company D was concerned about OSHA regulations because it had a situation in which humans were bending at the waist and reaching into plastic injection modeling machines to unload parts. This procedure required special safety equipment which impeded the workers. Company E placed robots in jobs requiring the loading of hot forged metal pieces into a trim press. Again, the company would be relieved from concerns about worker safety and accompanying safety equipment.

Comparison of Expected and Actual Results of Robot Employment

This comparison is on a company by company basis. The firms did not assign specific dollar values, unit amount, or percentages of savings and/or increases in any of the areas of expected results. Those individuals interviewed did indicate that they expected the robots to perform to the specifications provided by the vendor. Robot vendors typically phrase expected results in terms of an expected 20 to 30 percent increase in output. Some firms were able to quantify operating results after collecting data. The expected results are listed in the sequence of most to least important for each company.

Table 5.3 Company A

Expected Results	Actual Results
1. Reduce labor costs	Yes, by 30%
2. Increase output	Yes, by 20%
3. Relieve worker shortage	Yes
4. Relieve tedious jobs	Yes
5. Reduce capital costs	Not evaluated
6. Improve quality	Yes*
7. Remedy turnover	No change
8. Decrease waste	Yes, from 10% to 0.5%
9. Increase flexibility	Not evaluated
10. Ease OSHA compliance	No change

*This is reflected in the "Decrease waste" category.

The results of employing robots at Company A were in line with expectations. The impact on labor costs, output, and waste make this installation appear to be close to the ideal insofar as positive results are concerned. The above results will be explored further under the areas of impact on workforce, methods of financing robotics, future plans for employing robots, and strategic implications for Company A and the remaining companies, as well in the sections of this chapter which follow the comparison with the Japanese study.

Like Company A, Company B had impressive results in the areas of labor cost reduction and increased output. Company B did not significantly decrease waste. The reason was that products had been reworked instead of discarded, hence some of the quality

Table 5.4 Company B

	Expected Results	Actual Results
1.	Reduce labor costs	Yes, 30 to 50%
2.	Improve quality	Yes
3.	Increase output	Yes, 2 to 10 times*
4.	Relieve tedious jobs	Yes
5.	Decrease waste	Yes, 1%
6.	Increase flexibility	Yes**
7.	Reduce capital costs	Yes, by 2%
8.	Remedy turnover	Yes
9.	Relieve worker shortage	Yes
10.	Ease OSHA compliance	Not evaluated

*Depends upon product being fabricated.
**Allows modular fixturing and group tooling.

improvement credited to the robots shows up as increased productivity because of the negligible amount of rework now required. Company B was the only company to identify a capital cost reduction. Company D feels that dedication of its robots to one particular product will not generate a savings in capital costs. Companies A, C, and E had not addressed this issue.

Table 5.5 Company C

	Expected Results	Actual Results
1.	Increase output	Yes
2.	Improve quality	Yes
3.	Increase flexibility	Yes
4.	Decrease waste	Yes
5.	Relieve tedious work	Yes
6.	Remedy turnover	Yes
7.	Relieve worker shortage	Yes
8.	Reduce capital costs	Not evaluated
9.	Reduce labor costs	Yes
10.	Ease OSHA compliance	Yes

The assembly robot installed by Company C has not been evaluated numerically in terms specifically identifiable as the expected results listed above. The company management purchased the robot because it is able to consistently apply a silicon glaze to windows. Consistent performance from humans in this task was not possible because of the properties of the glazing compound. The company is satisfied that the robot performs as anticipated. The firm's satisfaction with the initial robot installation led to the decision to use welding robots and to investigate other areas for potential robot employment.

Table 5.6 Company D

Expected Results*	Actual Results
1. Reduce labor costs	Yes, by 50%
2. Improve quality	Yes, 10% per shift
3. Increase output	No
4. Decrease waste	Yes, by 5%
5. Increase flexibility	No, decrease
6. Ease OSHA compliance	Yes
7. Relieve tedious jobs	Yes
8. Relieve labor shortage	Not evaluated
9. Remedy turnover	Not evaluated
10. Reduce capital costs	No, increased**

*The information above is for the load/unload robots in the molding department. The assembly robots have not been in place long enough for evaluation, but are operating satisfactorily and the anticipated payback period is one year on the $150,000 investment.

**Since the robots are dedicated to one task and the company does not foresee other uses, the investment is not seen as one which could reduce costs or save future costs in other production areas.

Company D takes a singular and specific stance toward the employment of robots. That is, the primary justification for robotics is an economic trade-off in human man-hours saved by installing robots. Factors other than reducing labor costs are not given priority in the decision-making process.

Table 5.7 Company E

Expected Results	Actual Results*
1. Relieve tedious jobs	Yes
2. Improve quality	Yes +
3. Increase flexibility	Yes
4. Decrease waste	No
5. Ease OSHA compliance	Yes
6. Relieve worker shortage	Yes
	Reduced labor turnover*

+ Although not ranked as the number one reason for employing robots, the company stated that a desire to tighten specifications on the product was the reason for considering robots for the job.

* Company E ranked only six expected results from employing robots. However, a reduction in labor turnover, which was not anticipated, occurred.

At the time data was collected, one would conclude that Company E had not achieved a successful robotics implementation. The reduction of labor turnover was an unanticipated benefit of the robotics installation at Company E. In spite of technical problems, the firm expressed an intent to pursue the robotics installation to a successful conclusion.

Comparison of Findings in this Study and Findings of a Japanese Study

A study on the employment of robots in Japan [55] includes findings for large and small firms and the findings in both categories are parallel. The reasons for employing robots found in this study are compared with similar findings from the Japanese study in Table 5.8:

Table 5.8 Comparison from Two Studies of Reasons for Employing Robots

Factor	Rank This Study	Rank Japanese Study
Increase output	1	1
Improve quality	2	2
Reduce labor costs	3	4,7*
Relieve tedious jobs	4	10
Decrease waste	5	3+
Increase flexibility	6	5
Relieve skilled worker shortage	7	6
Remedy labor turnover	8	9
Reduce capital costs	9	3+
Ease OSHA compliance	10	NA

* While the Americans seem to think of labor savings in terms of reducing labor costs, the Japanese take a two pronged approach to this matter. One aspect is the actual reduction of the amount of labor required to produce something. This is identified by the Japanese as a factor separate from reducing labor costs. Reducing labor was ranked number 4 by the Japanese, and reducing labor costs was ranked number 7.

+ The term "rationalize production processes," is used by the Japanese and involves methods of scheduling and arranging work processes so as to get the greatest possible productivity from available resources. This process encompasses decreasing waste and reducing capital costs. The decreasing of capital costs and decreasing waste are not addressed as specific items in the Japanese study.

Results of Employing Robots

Overall, there appear to be no major differences between the expected and actual results of employing robots in those firms examined in this study. The same statement applies to the results of the Japanese study. Additionally, review of the U.S. firms studied reveal that some results in the previous tables translate into factors not considered in the initial evaluations, but were identified in answers in the questionnaire and fol-

Table 5.9 Comparison from Two Studies of the Results of Employing Robots

Results*	U.S.	Japan
1. Improved quality (precision)	Yes	Yes
2. Enhanced productivity	Yes	Yes
3. Reduced labor costs	Yes	Yes
4. Improved worker skill levels	Yes	Yes
5. Improved working conditions	Yes	Yes
6. Easier, more consistent production scheduling	Yes	Yes
7. Stabilized production	Yes	Yes
8. Improved worker attitudes	Yes	Yes
9. Relieved skilled labor shortage	Yes	Yes
10. Increased product reliability/fewer quality control inspections	Yes	Yes
11. Improved safety/fewer accidents	Yes	Yes
12. Solved labor shortage problems	Yes	Yes
13. Shorter setup times/flexibility	Yes	Yes
14. Product diversification	+	Yes
15. Entered new markets	Yes	Yes
16. Decreased dependence upon suppliers	Yes	Yes
17. More confidence from customers	Yes	Yes
18. Reduced equipment costs	Yes	Yes
19. Labor savings	Yes	Yes
20. Decreased waste/scrappage	Yes	+ +
21. Relieve workers from tedious/dangerous jobs	Yes	Yes
22. Reduced stock on hand	Yes	Yes
23. Better able to meet delivery schedules	Yes	Yes
24. Skill levels of operators and maintenance personnel increased	Yes	Yes

* Not all results apply to each firm.

+ Though product diversification was addressed by the U.S. firms, this has not happened. The Japanese study reported some Japanese firms had diversified as a result of employing robots.

Source: Robotization: Its Implications for Management, JMA Research Institute, Toyko, Japan, 1983, p. 266–276 and this study.

low-up interviews. The following list is a composite of the results and includes factors which are by-products of the original reasons for and results of employing robots. Consequently more than 10 results are identified in this list in contrast with only 10 factors in the preceding tables.

User Concerns About Employing Robots

In all five cases presented in this study, there were no significant objections encountered to the employment of robots in the plants. Objections which were raised by the incumbent workers related to job security. When management explained that no jobs would be lost and employees would have an opportunity to train for jobs related to operating and maintaining the robots, the employees willingly accepted them. Only one firm took what could be termed drastic steps to protect the robot from possible sabotage by employees. That firm placed the robot inside a fenced area with restricted access. Management soon realized that this action was unnecessary and the fence was removed within six months after it was installed.

Strategic Implications of Employing Robots

The strategic implications of the employment of robotics are those related factors which have the potential for affecting the long-term operations and providing a competitive advantage to the firm. In reviewing the above results of employing robots the following factors were considered of strategic significance for the management of the firms studied.

1. Productivity enhancement through the increasing of output rates, reducing costs, decreasing scrap, and increasing quality provide the firm the opportunity to:

a. increase market share by offering higher quality at the same price or lower a price.

b. increase profitability through lower service and/or warranty costs.

c. increase profitability by offering goods with lower production costs at competitive prices.

d. enter new markets by becoming a supplier of parts which were previously purchased from others.

e. decreasing dependence upon outside suppliers.

2. Decreased setup times and increased flexibility place the firm in a position to respond to changing market demands and meet delivery schedules providing opportunity for expanding markets and/or entering new markets with new product lines.

3. Reducing long-term capital costs creates an opportunity for alternative uses of capital funds.

4. The stable workforce represented by the robots plus the alleviation of skill shortages and reduction of labor turnover allows for confidence in long-term planning.

5. The employment of robotics could act as an entry barrier in certain industries. For example, competitors in the manufacturing of metal doors and windows consider anyone with a sheet metal fabricating capability to be a potential entrant into the market. Company C felt that the quality produced through the use of robots could deter those whose handmade products would be of lower quality and possibly higher cost.

6. The employment of robots can be the first step toward more sophisticated automated manufacturing systems which could amplify those strategic implications mentioned above.
7. Robotics can enhance company image. This idea was illustrated by Company B's prominent featuring of its advanced manufacturing technology (robots) in company brochures and professional journal advertisements.
8. Robotics can have a significant impact on those manufacturing processes which directly or indirectly interface with the automated part of the manufacturing process. For example, Company B found it necessary to increase efforts in those processes supplying parts to the robots because of the consistent pace of the robots' output. Areas such as purchasing stock, work in process inventory, finished goods inventory, and quality control activities can also be impacted.
9. Unfortunately, some situations in implementing robotics can have negative effects in both the short and long term. Companies C and E had difficulty in placing the robots (welding-Company C) (loading/unloading-Company E) into operation. Apparently these companies agreed to be test sites for technology not adequately proven by the manufacturer. Such experiences could cause a firm to cease pursuing possible advantages of advanced manufacturing technologies. Firms should ascertain whether they are dealing with a proven technology.
10. Although one of the firms visited did not participate in the study, an observation at that firm could be applicable to any firm consider-

ing the employment of robotics. No replacement for a failed control circuit was available from the manufacturer. This caused the firm to revert to manual operation for about four weeks. A firm should examine contingency requirements in terms of service, spare part support, and ability to maintain production through alternative methods should a robot system fail.

Future Plans Concerning Robots

Although specific plans had not been made insofar as numbers and types of robots that would be considered for future employment, all firms except Company E indicated the intent to use more robots in the future. Company E will evaluate its current position before making a decision. Companies B and C have established management positions with the specific responsibility of investigating robotics and other advanced manufacturing technologies for future consideration.

Other Results

Impact on the Workforce

Besides the workforce impacts indicated in Table 5.9, other impacts in some firms deserve attention. In all firms, personnel were trained to become programmers, robot maintainers, or robot operators. As will be seen in the comments on the specific firms which follow, some impacts were quite dramatic, others more routine.

Company A.

This firm experienced the most dramatic impact from employing robots when compared with the others. Company A increased its workforce from 48 to 68 people and added a second shift because of the added capabilities presented by the robots. New positions such as robot operator or programmer were filled by training incumbents. No new management positions were created.

Company B.

The employment of robots in this firm created 20 new positions for the hourly workforce. These positions were filled with 2 new hires and 18 with retrained incumbents. A new department, the Robotics Department, with 2 positions was created. The positions involved are titled robotics engineer and robotics supervisor and were filled by incumbent personnel. The functions of the department are to determine the most appropriate use of installed robotics and to investigate future employment of robotics in the plant.

Company C.

Two new departments and three new management positions were created at this firm as the result of employing robotics. The departments are Advanced Technology, headed by the senior analyst advanced technologies, and manufacturing engineering, headed by the chief manufacturing engineer. The other management position is in the Advanced Technology Department. While the above positions deal with technologies other than robotics, two new robot specific positions were created. They were filled by training incumbents and are called lead operators. These are considered management positions. In the hourly workforce, 4 new positions were created. All were filled by retraining incumbent personnel.

Company D.

No new personnel were hired because of robots and no one lost a job; rather, there was a reassignment of tasks. One operator and one quality control inspector from each of three shifts, were assigned other jobs in the plant. The quality control function is now performed by the one operator assigned to each robot on each shift.

Company E.

As in Company D, no new personnel were hired after robots were installed. Robot related positions were filled by retraining incumbents.

Methods of Financing Robots

Each of the five firms studied used different financing techniques. These were:

Company A—Paid cash from internally generated funds.

Company B—Financing through municipal revenue bonds.

Company C—Third party lease.

Company D—Bank financing.

Company E—Financing through parent company.

The above information indicates that many financing alternatives are available to the firm considering installing robots. Other sources of financing include direct lease from the producer and financing services offered by some larger robot producers. Companies A and C indicated they felt that financing was not an impediment to the adoption of advanced manufacturing technology. The other three firms thought financing was a problem for the small manufacturer.

Approaches to Robot Installation Decisions

In each case the final decision to employ robots was made by a team of company managers with each contributing some expertise. The engineering, installation, and training activities were carried out by the producers of the robots in conjunction with their distributors in four cases; in the other case, a robotics system house designed the system. The minor problems encountered in bringing the systems up to operating expectations were handled efficiently by the firms installing the robots. Companies C and E are undergoing more prolonged engineering changes for their robots, but do feel that solutions are near. However, two factors were brought out by managers in three of the firms which are important in the early decision stages. These are:

(1) the robotics systems may be so much more efficient than the supporting manufacturing processes that the other processes are stressed by robots;

(2) in one case the firm requested price quotes on the robot and did not specifically address issues such as computer control systems, engineering, and installation. The projected budget was based on the cost of the robot and its end device alone. This was about 50 percent of the actual installed cost.

Decision factors will be discussed further in Chapter 6.

Summary

This chapter has presented a descriptive analysis of the expected and actual results of employing robots in five small manufacturing firms located in the southeastern United States, as well as compared the findings with the findings of a study of the employment of robots in Japan. In addition, the following topics were specifically addressed: user concerns in employing robots, strategic implications of employing robots in the small manufacturing firm, future plans for the employment of robots by those firms studied, the impact of robotics on the firms' workforce, and methods of financing the installation of the robots.

It was found that the order of importance of reasons for employing robots or expected results was fairly consistent across the five firms studied. A comparison of similar factors from this study with the findings of a Japanese study revealed that both U.S. and Japanese manufacturers expected similar results from robotics.

Using robots also demonstrated consistent results across four of the five firms. The fifth firm has not resolved all its problems with robots, but feels benefits have been obtained and will continue to use robots. The fifth firm is also the only firm which indicated its interest in robotics was generated by a truly long-term concern, that of having the flexibility to manufacture products which would replace products facing obsolescence in 10 to 15 years. It is important to note that the number of identifiable results of employing robots was over twice the number of factors considered in the study. Since the reasons for employing robots identified in the study were typical, as identified by the robotics industry, it is apparent that robots are being sold as devices to help cure short-term problems without adequate emphasis on long-term benefits. For example, improved quality, increased output rate, decrease in scrappage, and the solving of labor shortages have resulted in the ability of some firms to enter new markets, expand current markets, and plan and control production in a more effective manner than had been possible before.

No negative impacts on the workforce resulted from employing robots. In fact, no workers lost jobs, skills were upgraded, and in one case the workforce was increased by 30 percent. Also, changes in organizational structure took place in which new departments and management positions were created to take better advantage of the potential of robotics.

It is significant that four of the five firms studied intend to increase use of robots in the future even though specific applications and types had not been identified at the time of the research. The fifth firm has not ruled out the purchase of additional robots, but will study results further before making a decision.

As mentioned above, results were identified which came from a combination of the expected results. It is these factors which generated the strategic implications of employing robotics on the part of the small manufacturer. For example, Company A did not envision itself as a supplier of parts to the competition when it installed robots, but this was an end result. Additionally, the firm was better able to compete in the international market and won a sizeable contract in bidding against six foreign competitors, including Japan. Other companies have increased profitability from operations and are more aggressively pursuing new markets. Robotics is seen as an entry barrier to future competitors who cannot produce as high a quality and at as high a rate through nonautomated production methods. These features create an advantage over competitors who have not pursued automation. Savings in projected capital equipment costs have freed funds for alternative investments or pursuit of additional automation. Also, recognition of the beneficial results gained from robotics has led firms to create new departments and management positions to guide companies in future automation efforts as well as management of current robotics installation.

The data and analyses presented in this chapter indicate that positive results are possible and should be expected from a properly planned and executed robotics installation.

CHAPTER 6

Conclusions and Recommendations

This chapter will present a summary of the methodology employed in the conduct of the study, limitations of the study, summary of the conclusions, implications of the study, and recommendations for future research.

Summary of the Methodology

This study was conducted through field research and a review of pertinent literature. The researcher visited seven manufacturing plants in the southeastern United States. Five firms agreed to participate in the study and at each plant appropriate management personnel completed a questionnaire designed to provide the basis of a case study for each field site. The questionnaire was supplemented by onsite interviews and observations from tours of each plant. Where further clarification or amplification of information obtained during the visits was required, follow-up was via telephone. From the data gathered during the field research, a brief case study was developed for each firm visited. The case described the firm's operations, reasons for using robots, results of using robots, concerns in employing robots, strategic implications of using robots, method of financing the robot installations, and future plans concerning the use of robots.

The case study data was analyzed to identify strategic implications of employing robots in small manufacturing firms. This was accomplished through analyzing expected results versus actual results of employing robots by the subject firms both on an individual firm basis as well as across firms in an attempt to identify common expectations and results from employing robots. The findings of this study were compared to the findings of a similar Japanese study.

Limitations of the Study

This study had several limitations. One was that very few manufacturing firms in the category of 2,000 or fewer employees per enterprise located in the southeastern United States employ robots. Only nine potential research sites were identified, and five of these agreed to cooperate in the study. The primary reason behind the low robotics use is probably because robotics is a relatively new phenomenon in the United States, with most robots being utilized by large firms such as automobile manufacturers. Also, robotics producers are concentrating their sales efforts on the large manufacturing firms in an attempt to obtain large orders to recoup their startup costs. The concern about being able to generalize from the small sample size to the population of small manufacturing firms was somewhat alleviated by the consistency of the findings from the collected data and the common factors found when findings in this study were compared with the results of the Japanese study. Unfortunately, there is not a case study for a firm which tried robots, but discontinued their use because of unsatisfactory results. One such firm was located, but the owner declined to participate in the study.

One limitation to the analysis of data was that the firms would not disclose figures on the financial aspects of their operations except for information such as gross revenues. While more financial information could have been beneficial to analysis, the primary purpose of the research was served in that the data obtained did reveal some strategic implications for the employment of robotics in the smaller manufacturing firms. Another limiting factor was a lack of travel funds, hence the restriction of field visits to one geographic region. The availability of more travel funds would have enabled the researcher to collect more information for the study.

A final limitation lies in the research instrument. The necessity to construct a questionnaire brief and concise enough to elicit the cooperation of busy managers, even when reinforced by interviews, tends to set the tone for more brevity in the data provided than may be desirable for a study with the strategic implications identified thus far.

Summary of the Conclusions

The findings from analysis of the data in Chapter 5 addressed the five research questions posed initially in Chapter 1. Principal reasons for employing robots, user concerns, the results of employing robots, strategic implications of employing robots, and the future plans for employing robots the firms studied were identified. Additionally, other relevant information was analyzed from the data collected during the field research and compared with the findings of this study with those of a Japanese study.

This study provides evidence that small manufacturing firms can achieve positive results by employing robots in their manufacturing processes. The analysis of expected versus actual results of utilizing robotics indicates that the typical approaches used in the decision-making process concerning robots focus on individual short-term results and not the synergistic effects that may be obtained through exploiting combinations of the expected results.

The findings in this study are supported by findings in large firms in the United States which employ robots and of both large and small firms which employ robots in Japan. The parallels in findings of this study and the Japanese study are significant because there are several times as many robots utilized in manufacturing processes in Japan as there are in the United States.

For the small manufacturing firm, robotics may play a more significant role in relieving the shortage of skilled workers than in the large firms. Although difficulty in the financing of automation projects is often raised as a reason for the small manufacturer not to pursue automation, this study reveals that many financing alternatives are available which should be investigated.

Those small firms using robots have found increased productivity from the application of a stand-alone type automation such as robotics which can interface with present production systems. This is true because robots can take over jobs that are dangerous, tedious, and difficult for humans while humans can perform other jobs in the production processes.

The producers and distributors of robots can play a more significant role in the introduction of robotics to the small manufacturing firms if they would recognize that the smaller firms do represent a market which could be developed. Given the extensive use of robots by small firms in Japan, if U.S. producers do not become active in this area, another domestic market will be lost to imports as the Japanese intensify efforts to export robots to the United States in the same manner that they have done with machine tools and other manufacturing technologies.

The findings that two firms did not fully anticipate the stress the increased productivity of the robots could place on the points of interface between the robots and the remainder of the system, plus the fact that in one case the pricing system was not fully understood, indicate a need for the education of small manufacturers on the decision areas to be explored when considering automation.

Concerns about the negative impact robots might have on job security were dispelled through the combination of more jobs becoming available and retraining of workers in the firms employing robots. Additionally, the skill levels of incumbent workers have been raised.

Implications of the Study

The 24 results from the employment of robotics which were identified in Chapter 5 indicate that there are significant strategic implications which argue strongly for efforts to accelerate the introduction of robotics to the small manufacturing firms in the United States. Strategic results such as enhanced productivity, increased quality, expanded markets, entry into new markets, and freeing of capital funds for alternative investments by the firms studied provide examples of forward thinking which can be applied to many small manufacturing firms feeling pressures from foreign goods in domestic and overseas markets.

The strategic implications are that in the long term U.S. firms which invest in advanced manufacturing technology may gain market share from both offshore producers and foreign producers who depend upon lower wage rates in foreign countries for their competitive edge. This could be particularly true if American citizens respond to the "buy American" campaigns being conducted by U.S. industry.

Implications for Government Agencies

A number of agencies exist who could play positive roles in helping improve the competitive position of the United States by assisting in the early transfer of robotics to the small manufacturers. The Congress of the United States in tax reform measures could provide specific investment incentives which would benefit the small manufacturers to the same degree as large firms. Through currently funded research activities the

National Science Foundation and the Office of Technology Assessment could provide programs to educate potential users about the benefits of employing robots.

Implications for Robotics Manufacturers/Distributors

Robot producers themselves should take a prime role in education. Each year trade shows and exhibitions demonstrate the latest manufacturing technology. However, these activities are typically applications oriented and are of more interest to technical and engineering people than to general management. Some trade shows are beginning to have seminars on considerations about the employment of robotics. Unfortunately, most of the educational effort focuses on short-term economic analyses with little emphasis on the synergistic effects of a strategic nature. Other manufacturing and trade organizations should be participants in the education effort.

Implications for Educational Institutions

Educational institutions are in an excellent position to provide information to local manufacturers on how to go about identifying and evaluating the decision factors necessary to help ensure successful implementation of robotics and/or other advanced manufacturing technologies. Those colleges and universities which have small business institutes or small business development centers should be especially alert to ensure that both established and proposed businesses are aware of the potential benefits robotics can offer.

Implications for Financial Institutions

Another sector which could play an active role in the education process is the financial sector. This sector should be involved both as student and teacher.

Financial institutions should be informed of the benefits of gaining expertise to help small manufacturers analyze needs, costs, and benefits of technologies for which they seek financing. The financial institutions can be more aggressive in seeking business from those small manufacturers who show that they can think in the long term.

Revised Decision Model

One educational device which could be of benefit to the small manufacturer investigating the employment of robotics is the decision model presented in Figure 6.1. An initial model was developed and presented in Chapter 2. The basic structure of the model has not changed. However, the findings in the study indicate the need to revise the model to include additional factors in both the Strategic and Tactical Factors areas. Under Strategic Factors, a new category called "New Markets/Products." Company A was surprised to find a new market due to its use of robots. However, it is recommended that this area be explored early in the decision process. From the strategic standpoint, the model can allow a firm to identify and target new markets and/or identify potential new products.

New Tactical Factors are "React to Market Demands" and "Production Process Impact." The first refers to the flexibility which allows the firm to produce more of a part (Company A for example) when market demand changes, without upsetting overall production activity and scheduling. Company A and Company B provide examples of the need to consider the impact on the overall production process of robotics. Company A added a new shift and Company B had to increase capacity for those processes which are paced by the robot. The additions are indicated by an asterisk in Figure 6.1, Revised Decision Model for Robotics Adoption. The model identifies some important decision factors which should be considered by

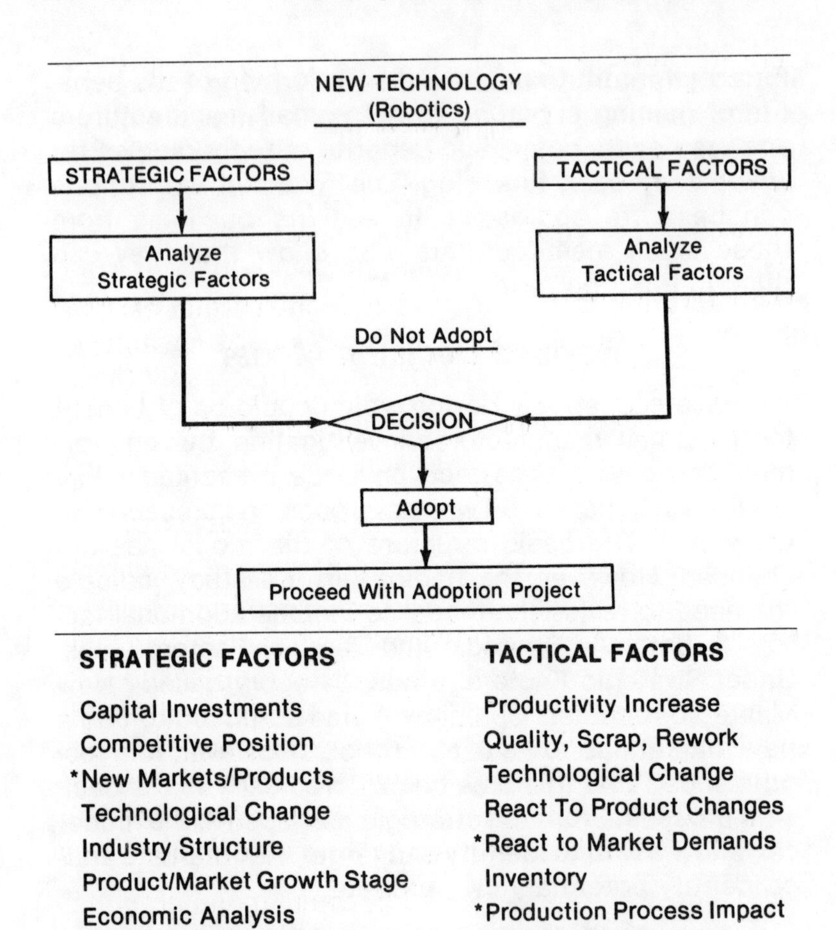

STRATEGIC FACTORS

Capital Investments
Competitive Position
*New Markets/Products
Technological Change
Industry Structure
Product/Market Growth Stage
Economic Analysis
Organizational Structure
Demographics
Political-Legal Factors
Government Policy
Others As Applicable

TACTICAL FACTORS

Productivity Increase
Quality, Scrap, Rework
Technological Change
React To Product Changes
*React to Market Demands
Inventory
*Production Process Impact
Economic Analysis Cost
Work Force-Training
OSHA Requirements
Others As Applicable

Fig. 6.1. Revised Decision Model For Robotics Adoption.

those contemplating the adoption of robotics [5]. Obviously, the tactical factors are also the previously identified results which provide the synergy to transform what may seem to be short term into the strategic implications identified in this study and support the strategic factors to be considered. Training in the use of models such as this can provide the small manufacturer with a rational approach to the robotics adoption decision process.

While this study identifies the strategic role robotics can play in the competitive position of U.S. manufacturing firms, it is also recognized that certain operations require unique human thought capacities that are not subject to automation at the present time.

Recommendations for Future Research

Spokespersons for the robotics industry indicate that the major reason (besides small order quantities) they do not pursue sales to small manufacturers is that responses to queries from them often require site visits and engineering studies which are expensive and small manufacturers do not understand the need to pay for such services. Thus the return on the investment of the producer's or distributor's time is negligible [129]. Research is needed to help determine:

1. which processes used by small manufacturers are most amenable to automation;
2. the degree of knowledge possessed by small manufacturers about robotics in terms of both the immediate and the long-term benefits which are possible through using robots;
3. the desire of small manufacturers to adopt robotics;
4. the most appropriate and efficient methods of educating small manufacturers on the potential benefits of employing robots;
5. the roles that should be played in accelerating the transfer of robotics to small manufacturers.

These research activities will require efforts on the part of a number of agencies. Some agencies and beneficial research roles they might play are:

Government Agencies

Research is needed which will identify positive and practical roles the government can play in assisting the transfer of robotics to the small manufacturer. Areas for consideration are investment incentives such as special depreciation allowances aimed specifically at the small manufacturer, guaranteed loans, increased availability of revenue bonds for automation projects, assistance in developing overseas markets, and providing automation assistance to small manufacturing firms that bid on government contracts.

Trade Organizations

Trade organizations such as the Robotics Institute of America can revise their data collection and reporting procedures to identify users of robots in the small manufacturer category. This information can provide contact points for small manufacturers who are interested in investigating the use of robots. Currently, such information is not available. For example, during the research for this study it was discovered that three small manufacturers in one metropolitan area employed robots. However, none knew that the others had robots.

Educational Institutions

Education institutions with robotics research facilities should find methods to provide information specifically to small manufacturers in their areas of service. In this vein, those institutions with small business programs should undertake research aimed at determining interest in and need for automation in their fields. Educational institutions should study small manufacturers and strategic operations based on the Wheelwright model presented in Chapter 2. Studies of this nature could prove fruitful in helping identify differences between and similarities in strategic management of large and small firms.

Professional Organizations

Professional organizations such as the American Manufacturing Association and the American Production and Inventory Control Society should pursue research projects which identify the needs of small manufacturing special interest groups and provide an information clearinghouse for members.

Although the research recommendations above overlap in terms of having several organizations target small manufacturers for research and education purposes, it should be remembered that there are over 300,000 manufacturing firms in the United States with fewer than 2,000 employees. Therefore, there should be no concern for "overkill" in the research efforts.

Finally, more research efforts such as this study should be undertaken to assemble a broader data base on the strategic implications and results of the employment of robotics by small manufacturing firms. Especially helpful would be preadoption studies followed by postadoption studies with firms installing robotics. Such a study would be useful in the assessment of this study and would provide a basis for longitudinal studies.

APPENDIX A

The Research Instrument

Questionnaire on the Employment of Robotics by the Small Manufacturer

INSTRUCTIONS: This questionnaire is designed to obtain information about your firm's decision to employ robots in the manufacturing processes. The purposes of collecting the information are to help the researcher complete his Ph.D. in Business Administration at the University of Georgia and to provide information to the small business community and the robotics producers about the important role robotics can play in the small manufacturing firm. If desired, the identity of your firm will not be revealed in the data analysis or the publication of the results. Since little data has been collected to date on the use of robots by small manufacturers, your assistance will be a significant contribution. The questionnaire can be completed in about twenty minutes.

I. Company Profile

1. Your company's industry in which robots are used?
   ____

2. Title of individual completing this questionnaire.
   ____

3. The annual sales volume of your company for
   1981:____________________ 1982:____________________
   1983:____________________ 1984:____________________

4. The number of employees in your firm in
   1981:____________________ 1982:____________________
   1983:____________________ 1984:____________________

5. Which term best describes your firm (place % on line)?
   _____% make-to-stock
   _____% make-to-order
   _____% other (please specify)______________________.

6. Type of process _____ Job shop _____ Continuous process
   _____ Assembly line.

7. Approximate number of "end items" in master schedule (excluding service parts): ________________

8. Approximate number of different part numbers: ________

9. Number of levels in bill of materials________________

10. The production planning and control system is
_____ manual _____ computerized.

II. Types of Robots and Applications

11. In what year did you first introduce robots into your plant?
____

12. How many robots were in use in your plant in Jan 1980? _____,
Jan 1981? _____, Jan 1982?_____, Jan 1983? _____,
Jan 1984? _____.

13. How many robots are now being used in each of the following uses?

Number	*Manufacturer/Type*	*Cost(complete)*
____spot welding	__________	__________
____spray painting/ coating	__________	__________
____loaders/ unloaders	__________	__________
____assembly	__________	__________
____continuous seam welding	__________	__________
____material handlers	__________	__________
____other (specify)	__________	__________

If more space is needed, write data below.

14. Robots were placed in a ____ new plant, ____ old plant,
____ both?

III. Strategic Factors Considered in Employing Robots

15. Who initiated the interest of your firm in robotics?

16. What factors or events initiated the interest?

17. Who was involved in making the decision to employ robots (position/function)?

18. What were their roles in the decision process?

IV. Principal Reasons for Employing Robots

19. Please rank the following items in order of importance as reasons for the use of robots. Most important will rank number 1. Use N/A for any factor which was not considered.
    ____ to reduce labor costs
    ____ to ease compliance with OSHA regulations
    ____ to reduce capital costs
    ____ to improve product quality
    ____ to increase output rate
    ____ to increase the flexibility in changing products, designs, etc.
    ____ to decrease material waste
    ____ to relieve workers of tedious and/or dangerous jobs
    ____ to remedy high labor turnover problems
    ____ to relieve shortages of skilled workers
    ____ other factors (please list)

V. User Concerns in Adopting Robots

20. How far in advance were operators made aware that robots would be introduced into their plants? ________months

21. How were operators made aware that robots would be placed in the plant?
    ____ through labor union ____ by meeting with management
    ____ were not notified prior to introduction of robot(s)
    ____ other (specify)

22. Were there objections to employing robots? ____ yes, ____ no

23. What were the objections?

24. Who objected? ____ managers, ____ hourly workers

25. How were the objections overcome?

26. Please describe the manner by which you financed the purchase of robots for your plant(s).

27. Do you feel that financing of capital equipment is a problem that is delaying the introduction of robotics and/or other advanced technology into the small manufacturers plant?

28. What recommendations do you have which might enhance the financing of advanced manufacturing technology for the small manufacturer? This can include sources of financing, taxation, government assistance, and so forth.

VI. Results of Employing Robots

29. Which of the following occurred after introducing robots?
____reduced labor costs by ________%
____eased compliance with OSHA regulations
____reduced capital costs by ____________________
____improved product quality, measure of improvement ______
____increased output rate by ________%
____decreased material waste by ________%
____relieved shortage of skilled labor
____relieved workers of tedious and/or dangerous jobs
____reduced labor turnover by ________%
____increased flexibility in changing products, designs, etc.
Examples of increased flexibility:

____other results:

30. Can you readily identify costs and savings attributable to using robots? Examples:

31. The introduction of robots has caused the following work schedule changes.
A. Average number of shifts per day: ____increased, ____same, ____decreased. Current number of shifts is ____, was____.

B. Average length of work shift per day: ____increased, ____decreased, __X__same. Length is _____, was _____.

32. Have quality control specifications been tightened? ________

33. The number of quality control employees has ____decreased, ____increased, ____not changed. Current number ____, previous ____.

34. Have robots had any of the following effects on production?
____given ability to manufacture more complex designs.
____increased efficiency in raw materials use. Measure used:

____reduced setup times by _______%
____enabled the production of new products.

35. How many new positions were created as the result of employing robots? _______

36. How were new positions filled? (Number of workers)
____retraining workers ____hiring new workers

37. If positions were eliminated, how many workers were
____transferred without retraining, ____same plant, ____another
____retrained for a new position within the plant
____retrained for a new position in another plant
____laid off ____retired early?

38. In retraining efforts, did the union(s) ____choose who would be retrained, ____support your efforts, but not select retrainees, ____oppose your efforts, ____not applicable.

39. Since the first use of robots, the number of production related accidents has: ____decreased, ____no change, ____increased.

40. Was it necessary to develop new worker safety measures? ____

41. How have robots affected the skill levels required in the following areas?
A. Operators: ____increased skill level, ____decreased, ____same.
B. Maintenance: ____increased, ____decreased, ____same.
C. Production Scheduling: ____increased, ____decreased, ____same.
D. Materials Handling: ____increased, ____decreased, ____same.

E. Other (specify):____
____increased, ____decreased, ____same.

42. Did the supplier of the robot help with training? __________

43. How long was the training period in days for
___maintenance, ____production scheduling, ____operators,
____other (specify) ________________________________

44. Have any new organizational units been created as a result of employing robots? _____ Title: ______________

45. How many departments were eliminated or consolidated? ____
Eliminated (identify) ____________________________
Consolidated (identify) __________________________

46. How has the use of robots affected the number of managers required? (number)
decreased by _____ increased by _____ no change _____

47. Have new robot-specific management positions been created? (Please specify)

48. If yes, how were the new positions staffed?
____retraining present managers
____promotion of workers ____new hires

49. To your knowledge have there been any personal problems caused by the introduction of robots? _____

50. Does your firm have a counselor to deal with worker problems caused by robots? (Full time staff member ___, part time ___)
Comments:

VII. Future Plans Concerning Robots

51. Do you plan to employ additional robots during the next five years? _______
____new plant, ____existing plant, ____both

52. If yes to the above question, has the number of use of robots been decided? _______
Please list types of robots and reasons for future use of robots as well as anticipated installation dates.

APPENDIX B
Data For Company A

Questionnaire on the Employment of Robotics by the Small Manufacturer

INSTRUCTIONS: This questionnaire is designed to obtain information about your firm's decision to employ robots in the manufacturing processes. The purposes of collecting the information are to help the researcher complete his Ph.D. in Business Administration at the University of Georgia and to provide information to the small business community and the robotics producers about the important role robotics can play in the small manufacturing firm. If desired, the identity of your firm will not be revealed in the data analysis or the publication of the results. Since little data has been collected to date on the use of robots by small manufacturers, your assistance will be a significant contribution. The questionnaire can be completed in about twenty minutes.

I. Company Profile

1. Your company's industry in which robots are used?
   Agricultural and Industrial Implements (Mowers)

2. Title of individual completing this questionnaire.
   Company Owner/President

3. The annual sales volume of your company for
   1981:______ 1982:______
   1983: $10 Million 1984: $12 Million

4. The number of employees in your firm in
   1981:______ 1982:______
   1983: 48 1984: 68

5. Which term best describes your firm (place % on line)?
   ______% make-to-stock
   100 % make-to-order
   ______% other (please specify)______.

6. Type of process X Job shop ______ Continuous process
   X Assembly line.

7. Approximate number of "end items" in master schedule (excluding service parts): ________________

8. Approximate number of different part numbers: __________

9. Number of levels in bill of materials________________

10. The production planning and control system is
__X__ manual __X__ computerized.

II. Types of Robots and Applications

11. In what year did you first introduce robots into your plant?
__1983__

12. How many robots were in use in your plant in Jan 1980? _____, Jan 1981? _____, Jan 1982?_____, Jan 1983? __4__, Jan 1984? __4__.

13. How many robots are now being used in each of the following uses?

Number	*Manufacturer/Type*	*Cost(complete)*
____spot welding		
____spray painting/ coating		
__4__loaders/ unloaders	GMF/Fanuc	3@$24,000, 1@$35K
____assembly		
__2__continuous seam welding	Installed April 1985, not operational as of 4/12. Shen-Meiwa $125,000 each	
____material handlers		
____other (specify)		

If more space is needed, write data below.

The welding robots are in checkout phase. We had wanted to purchase U.S. made robots, but could not find one which was satisfactory.

14. Robots were placed in a ____ new plant, __X__ old plant, ____ both?

III. Strategic Factors Considered in Employing Robots

15. Who initiated the interest of your firm in robotics?
Group effort by management.

16. What factors or events initiated the interest?

    More consistent production desired.

    Observed robots at the 1982 International Machine Tool Show in Chicago.

17. Who was involved in making the decision to employ robots (position/function)?

    Group effort by the President and Vice Presidents of this company. Function is Company Mgt.

18. What were their roles in the decision process?

    What types and brands of robots were to be used in what position.

IV. Principal Reasons for Employing Robots

19. Please rank the following items in order of importance as reasons for the use of robots. Most important will rank number 1. Use N/A for any factor which was not considered.

    __1__ to reduce labor costs
    __10__ to ease compliance with OSHA regulations
    __5__ to reduce capital costs
    __6__ to improve product quality
    __2__ to increase output rate
    __9__ to increase the flexibility in changing products, designs, etc.
    __8__ to decrease material waste
    __4__ to relieve workers of tedious and/or dangerous jobs
    __7__ to remedy high labor turnover problems
    __3__ to relieve shortages of skilled workers
    ____ other factors (please list)

V. User Concerns in Adopting Robots

20. How far in advance were operators made aware that robots would be introduced into their plants? __8__ months

21. How were operators made aware that robots would be placed in the plant?

    ____ through labor union __X__ by meeting with management
    ____ were not notified prior to introduction of robot(s)
    ____ other (specify)

22. Were there objections to employing robots? ____ yes, __X__ no

23. What were the objections?

24. Who objected? ___ managers, ___ hourly workers

25. How were the objections overcome?

26. Please describe the manner by which you financed the purchase of robots for your plant(s).

    Paid cash.

27. Do you feel that financing of capital equipment is a problem that is delaying the introduction of robotics and/or other advanced technology into the small manufacturers plant?

    Not in our case.

28. What recommendations do you have which might enhance the financing of advanced manufacturing technology for the small manufacturer? This can include sources of financing, taxation, government assistance, and so forth.

    Special tax and depreciation allowances should be granted to those who introduce new manufacturing technology.

VI. Results of Employing Robots

29. Which of the following occurred after introducing robots?

    _X_ reduced labor costs by _30_ %
    ___ eased compliance with OSHA regulations
    ___ reduced capital costs by ___
    ___ improved product quality, measure of improvement ___
    _X_ increased output rate by _20_ % per shift
    _X_ decreased material waste by _2,000_ %
    _X_ relieved shortage of skilled labor
    _X_ relieved workers of tedious and/or dangerous jobs
    ___ reduced labor turnover by ___ %
    _X_ increased flexibility in changing products, designs, etc.

    Examples of increased flexibility: Can now change part being produced in 15 minutes, it took one to days before.

    ___ other results:

30. Can you readily identify costs and savings attributable to using robots? _YES_ Examples:

    Exact production times can be figured to show production increase, less waste, less idle time.

31. The introduction of robots has caused the following work schedule changes.
    A. Average number of shifts per day: _X_increased, ___same, ___decreased. Current number of shifts is _2_, was_1_.
    B. Average length of work shift per day: ___increased, ___decreased, _X_same. Length is _8hrs_, was _8hrs_.

32. Have quality control specifications been tightened? _YES_

33. The number of quality control employees has ___decreased, ___increased, _X_not changed. Current number ___, previous ___.

34. Have robots had any of the following effects on production?
    _X_given ability to manufacture more complex designs.
    _X_increased efficiency in raw materials use. Measure used: Output and scrap.
    _X_reduced setup times by _____% (See above comment-flexibility.)
    ___enabled the production of new products.

35. How many new positions were created as the result of employing robots? _25_

36. How were new positions filled? (Number of workers)
    _5_retraining workers   _20_hiring new workers

37. If positions were eliminated, how many workers were
    ___transferred without retraining, ___same plant, ___another
    _X_retrained for a new position within the plant
    ___retrained for a new position in another plant
    ___laid off   ___retired early?

38. In retraining efforts, did the union(s) ___choose who would be retrained, ___support your efforts, but not select retrainees, ___oppose your efforts, _X_not applicable.

39. Since the first use of robots, the number of production related accidents has: _X_decreased, ___no change, ___increased.

40. Was it necessary to develop new worker safety measures? _YES_

41. How have robots affected the skill levels required in the following areas?
    A. Operators: _X_increased skill level, ___decreased, ___same.
    B. Maintenance: _X_increased, ___decreased, ___same.

C. Production Scheduling: ___increased, ___decreased, _X_same.
D. Materials Handling: ___increased, ___decreased, _X_same.
E. Other (specify):___
___increased, ___decreased, ___same.

42. Did the supplier of the robot help with training? _YES_

43. How long was the training period in days for _15_maintenance, ___production scheduling, _3_operators, ___other (specify) ___

44. Have any new organizational units been created as a result of employing robots? _NO_ Title: ___

45. How many departments were eliminated or consolidated? _NONE_
Eliminated (identify) ___
Consolidated (identify) ___

46. How has the use of robots affected the number of managers required? (number)
decreased by ___ increased by ___ no change _X_

47. Have new robot-specific management positions been created? (Please specify) No

48. If yes, how were the new positions staffed?
___retraining present managers
___promotion of workers ___new hires

49. To your knowledge have there been any personal problems caused by the introduction of robots? _No_

50. Does your firm have a counselor to deal with worker problems caused by robots? (Full time staff member _0_, part time _0_)
Comments: None Required

VII. Future Plans Concerning Robots

51. Do you plan to employ additional robots during the next five years? __Yes__
____new plant, ____existing plant, __X__both

52. If yes to the above question, has the number of use of robots been decided? __No__
Please list types of robots and reasons for future use of robots as well as anticipated installation dates.

APPENDIX B

Research Instruments with Data Entered

Data For Company B

Questionnaire on the Employment of Robotics by the Small Manufacturer

INSTRUCTIONS: This questionnaire is designed to obtain information about your firm's decision to employ robots in the manufacturing processes. The purposes of collecting the information are to help the researcher complete his Ph.D. in Business Administration at the University of Georgia and to provide information to the small business community and the robotics producers about the important role robotics can play in the small manufacturing firm. If desired, the identity of your firm will not be revealed in the data analysis or the publication of the results. Since little data has been collected to date on the use of robots by small manufacturers, your assistance will be a significant contribution. The questionnaire can be completed in about twenty minutes.

I. Company Profile

1. Your company's industry in which robots are used?
   Defense Industry

2. Title of individual completing this questionnaire.
   Production Engineer

3. The annual sales volume of your company for
   1981:________ 1982: Company would not re-
   1983:________ 1984: lease financial data.

4. The number of employees in your firm in
   1981:________ 1982:________
   1983: 425 Approx. 1984: 425

5. Which term best describes your firm (place % on line)?
   ____% make-to-stock
   100 % make-to-order
   ____% other (please specify)________________.

6. Type of process __X__ Job shop ____ Continuous process ____ Assembly line.

7. Approximate number of "end items" in master schedule (excluding service parts): ____________

8. Approximate number of different part numbers: ____________

9. Number of levels in bill of materials____________

10. The production planning and control system is
    __X__ manual ____ computerized.

II. Types of Robots and Applications

11. In what year did you first introduce robots into your plant?
    1983

12. How many robots were in use in your plant in Jan 1980? ____, Jan 1981? ____, Jan 1982?____, Jan 1983? __1__, Jan 1984? __2__.

13. How many robots are now being used in each of the following uses?

Number	*Manufacturer/Type*	*Cost(complete)*
___spot welding		
___spray painting/ coating		
___loaders/ unloaders		
___assembly		
2 continuous seam welding	ASEA-Irb-6 Cinc.Milacron T3-746	$90,000 $118,000
___material handlers		
___other (specify)		

If more space is needed, write data below.

The welding robots are in checkout phase. We had wanted to purchase U.S. made robots, but could not find one which was satisfactory.

14. Robots were placed in a ___ new plant, _X_ old plant, ___ both?

III. Strategic Factors Considered in Employing Robots

15. Who initiated the interest of your firm in robotics?
    Upper management and Production Personnel

16. What factors or events initiated the interest?
    Labor rates, production quantities.

17. Who was involved in making the decision to employ robots (position/function)?
    President, v.p. finance, Production superintendent, Manufacturing engineer.

18. What were their roles in the decision process?
    Present needs, financial justification.

IV. Principal Reasons for Employing Robots

19. Please rank the following items in order of importance as reasons for the use of robots. Most important will rank number 1. Use N/A for any factor which was not considered.
    _1_ to reduce labor costs
    _10_ to ease compliance with OSHA regulations
    _7_ to reduce capital costs
    _2_ to improve product quality
    _3_ to increase output rate
    _6_ to increase the flexibility in changing products, designs, etc.
    _5_ to decrease material waste
    _4_ to relieve workers of tedious and/or dangerous jobs
    _9_ to remedy high labor turnover problems
    _8_ to relieve shortages of skilled workers
    ____ other factors (please list)

V. User Concerns in Adopting Robots

20. How far in advance were operators made aware that robots would be introduced into their plants? _2_ months

21. How were operators made aware that robots would be placed in the plant?
    ____ through labor union _X_ by meeting with management
    ____ were not notified prior to introduction of robot(s)
    ____ other (specify)

22. Were there objections to employing robots? __X__ yes, ____ no

23. What were the objections?
Fear of job loss.

24. Who objected? ____ managers, __X__ hourly workers

25. How were the objections overcome?
Discussion with mgt.

26. Please describe the manner by which you financed the purchase of robots for your plant(s).
Systems purchased through Industrial Development Board bonds (municipal).

27. Do you feel that financing of capital equipment is a problem that is delaying the introduction of robotics and/or other advanced technology into the small manufacturers plant?
Yes

28. What recommendations do you have which might enhance the financing of advanced manufacturing technology for the small manufacturer? This can include sources of financing, taxation, government assistance, and so forth.
Private investors, traditional financing, and government agency support should be investigated.

VI. Results of Employing Robots

29. Which of the following occurred after introducing robots?
__X__ reduced labor costs by __30-50__ % depending upon product.
____ eased compliance with OSHA regulations
__X__ reduced capital costs by __2%__
__X__ improved product quality, measure of improvement
__rework__
__X__ increased output rate by __200-1,000__ %
__X__ decreased material waste by __1__ %
__X__ relieved shortage of skilled labor
__X__ relieved workers of tedious and/or dangerous jobs
__X__ reduced labor turnover by ____ %
__X__ increased flexibility in changing products, designs, etc.
Examples of increased flexibility:
Modular fixturing, "family" or group tooling.

_X_other results:

Robotic welding has improved our ability to meet rigid MIL STDS.

30. Can you readily identify costs and savings attributable to using robots? Examples:
    Yes. Better understanding of capabilities, better estimating, decreased production time.

31. The introduction of robots has caused the following work schedule changes. /robot dept only
    A. Average number of shifts per day: _X_increased, ___same, ___decreased. Current number of shifts is _2_, was_1_.
    B. Average length of work shift per day: ___increased, ___decreased, _X_same. Length is _8hrs_, was ____.

32. Have quality control specifications been tightened? _YES_

33. The number of quality control employees has ___decreased, _X_increased, ___not changed. Current number _6_, previous _4_.

34. Have robots had any of the following effects on production?
    _X_given ability to manufacture more complex designs.
    _X_increased efficiency in raw materials use. Measure used:

    _X_reduced setup times by _____%
    _X_enabled the production of new products.

35. How many new positions were created as the result of employing robots? _20_

36. How were new positions filled? (Number of workers)
    _18_retraining workers _2_hiring new workers

37. If positions were eliminated, how many workers were
    ___transferred without retraining, ___same plant, ___another
    _X_retrained for a new position within the plant (20-30)
    ___retrained for a new positiomin another plant
    ___laid off ___retired early?

38. In retraining efforts, did the union(s) ___choose who would be retrained, _X_support your efforts, but not select retrainees, ___oppose your efforts, ___not applicable.

39. Since the first use of robots, the number of production related accidents has: ___decreased, _X_no change, ___increased.

40. Was it necessary to develop new worker safety measures? _Yes_

41. How have robots affected the skill levels required in the following areas?
    A. Operators: _X_increased skill level, ___decreased, ___same.
    B. Maintenance: _X_increased, ___decreased, ___same.
    C. Production Scheduling: _X_increased, ___decreased, ___same.
    D. Materials Handling: _X_increased, ___decreased, ___same.
    E. Other (specify): _Manufacturing tooling_
    _X_increased, ___decreased, ___same.

42. Did the supplier of the robot help with training? _YES_

43. How long was the training period in days for _5_maintenance, _NA_production scheduling, _5_operators, _3_other (specify) _welders_

44. Have any new organizational units been created as a result of employing robots? _YES_ Title: _Robotics Dept._

45. How many departments were eliminated or consolidated? _NONE_
    Eliminated (identify) ___
    Consolidated (identify) ___

46. How has the use of robots affected the number of managers required? (number)
    decreased by ___ increased by _2_ no change ___

47. Have new robot-specific management positions been created? (Please specify) Yes, Robotic dept., supervisor and engineer.

48. If yes, how were the new positions staffed?
    _X_retraining present managers
    _1_promotion of workers _1_new hires

49. To your knowledge have there been any personal problems caused by the introduction of robots? _No_

50. Does your firm have a counselor to deal with worker problems caused by robots? (Full time staff member _No_, part time _No_)
    Comments:

VII. Future Plans Concerning Robots

51. Do you plan to employ additional robots during the next five years? __Yes__
____new plant, ____existing plant, __X__both

52. If yes to the above question, has the number of use of robots been decided? 2-4 possible FY '86.
Please list types of robots and reasons for future use of robots as well as anticipated installation dates.
Arc welding and painting robots are included in budget planning. Installation may be in FY '86.

APPENDIX B
Data For Company C

Questionnaire on the Employment of Robotics by the Small Manufacturer

INSTRUCTIONS: This questionnaire is designed to obtain information about your firm's decision to employ robots in the manufacturing processes. The purposes of collecting the information are to help the researcher complete his Ph.D. in Business Administration at the University of Georgia and to provide information to the small business community and the robotics producers about the important role robotics can play in the small manufacturing firm. If desired, the identity of your firm will not be revealed in the data analysis or the publication of the results. Since little data has been collected to date on the use of robots by small manufacturers, your assistance will be a significant contribution. The questionnaire can be completed in about twenty minutes.

I. Company Profile

1. Your company's industry in which robots are used?
   Construction—doors and windows

2. Title of individual completing this questionnaire.
   Senior Analyst, Advanced Technology

3. The annual sales volume of your company for
   1981:______ 1982: $60 Million
   1983: $100 Million 1984: $120 Million

4. The number of employees in your firm in
   1981:______ 1982:______
   1983:______ 1984: 1,500 in 3 plants

5. Which term best describes your firm (place % on line)?
   ____% make-to-stock
   100 % make-to-order
   ____% other (please specify)______.

6. Type of process ____ Job shop ____ Continuous process
   X Assembly line.

7. Approximate number of "end items" in master schedule (excluding service parts): 9000
8. Approximate number of different part numbers: 29000
9. Number of levels in bill of materials 3-10
10. The production planning and control system is
 ____ manual __X__ computerized.

II. Types of Robots and Applications

11. In what year did you first introduce robots into your plant?
 1984
12. How many robots were in use in your plant in Jan 1980? ____, Jan 1981? ____, Jan 1982? ____, Jan 1983? ____, Jan 1984? ____.
13. How many robots are now being used in each of the following uses?

Number		*Manufacturer/Type*	*Cost(complete)*
____	spot welding		
____	spray painting/ coating		
____	loaders/ unloaders		
1	assembly	RSI (Hybrid)	$275,000
2*	continuous seam welding	Devilbiss (Plasma-thermal arc)	$140,000
____	material handlers		
____	other (specify)		

If more space is needed, write data below.

*These robots were in checkout stage and were undergoing testing by the manufacturer. No operational data was available.

The welding and assembly robots were in different plants.

14. Robots were placed in a ____ new plant, __X__ old plant, ____ both?

III. Strategic Factors Considered in Employing Robots

15. Who initiated the interest of your firm in robotics?
 President and Executive Vice President

16. What factors or events initiated the interest?
    A requirement for improving production processes arose.

17. Who was involved in making the decision to employ robots (position/function)?
    Sr. Analyst, Advanced Technology

18. What were their roles in the decision process?
    Normal systems development methodology.

IV. Principal Reasons for Employing Robots

19. Please rank the following items in order of importance as reasons for the use of robots. Most important will rank number 1. Use N/A for any factor which was not considered.
    _9_ to reduce labor costs
    _10_ to ease compliance with OSHA regulations
    _8_ to reduce capital costs
    _2_ to improve product quality
    _1_ to increase output rate
    _3_ to increase the flexibility in changing products, designs, etc.
    _4_ to decrease material waste
    _5_ to relieve workers of tedious and/or dangerous jobs
    _6_ to remedy high labor turnover problems
    _7_ to relieve shortages of skilled workers
    ____ other factors (please list)

V. User Concerns in Adopting Robots

20. How far in advance were operators made aware that robots would be introduced into their plants? ____0____months

21. How were operators made aware that robots would be placed in the plant?
    ____ through labor union ____ by meeting with management
    _X_ were not notified prior to introduction of robot(s)
    ____ other (specify)

22. Were there objections to employing robots? ____ yes, _X_ no

23. What were the objections?

24. Who objected? ____ managers, ____ hourly workers

25. How were the objections overcome?

26. Please describe the manner by which you financed the purchase of robots for your plant(s).
   Third party lease

27. Do you feel that financing of capital equipment is a problem that is delaying the introduction of robotics and/or other advanced technology into the small manufacturers plant?
   No

28. What recommendations do you have which might enhance the financing of advanced manufacturing technology for the small manufacturer? This can include sources of financing, taxation, government assistance, and so forth.

VI. Results of Employing Robots

29. Which of the following occurred after introducing robots?
   ____reduced labor costs by ______%
   ____eased compliance with OSHA regulations
   ____reduced capital costs by ____________________
   ____improved product quality, measure of improvement ____________
   ____increased output rate by ________%
   ____decreased material waste by ______%
   ____relieved shortage of skilled labor
   ____relieved workers of tedious and/or dangerous jobs
   ____reduced labor turnover by ________%
   ____increased flexibility in changing products, designs, etc.
   Examples of increased flexibility:

   _X_other results: SEE COMMENTS AT END OF QUESTIONNAIRE.

30. Can you readily identify costs and savings attributable to using robots? _No_ Examples: See comments.

31. The introduction of robots has caused the following work schedule changes.
    A. Average number of shifts per day: ____increased, __X__same, ____decreased. Current number of shifts is __1__, was____.
    B. Average length of work shift per day: ____increased, ____decreased, __X__same. Length is _____, was _____.

32. Have quality control specifications been tightened? ___Yes___

33. The number of quality control employees has ____decreased, ____increased, __X__not changed. Current number ____, previous ____.

34. Have robots had any of the following effects on production?
    __No__given ability to manufacture more complex designs.
    __No__increased efficiency in raw materials use. Measure used:

    __No__reduced setup times by ________%
    __No__enabled the production of new products.

35. How many new positions were created as the result of employing robots? ___3 in window plant, 1 in door plant___

36. How were new positions filled? (Number of workers)
    __4__retraining workers ____hiring new workers

37. If positions were eliminated, how many workers were
    ____transferred without retraining, ____same plant, ____another
    ____retrained for a new position within the plant
    ____retrained for a new position in another plant
    ____laid off ____retired early?

38. In retraining efforts, did the union(s) ____choose who would be retrained, ____support your efforts, but not select retrainees, ____oppose your efforts, __X__not applicable.

39. Since the first use of robots, the number of production related accidents has: ____decreased, __X__no change, ____increased.

40. Was it necessary to develop new worker safety measures? __Yes__

41. How have robots affected the skill levels required in the following areas?
    A. Operators: __X__increased skill level, ____decreased, ____same.
    B. Maintenance: __X__increased, ____decreased, ____same.

C. Production Scheduling: ____increased, ____decreased, ____same.
D. Materials Handling: ____increased, ____decreased, __X__same.
E. Other (specify): Computer Programming
__X__increased, ____decreased, ____same.

42. Did the supplier of the robot help with training? Yes

43. How long was the training period in days for __½__maintenance, __½__production scheduling, __1__operators, __3__other (specify) Programmers

44. Have any new organizational units been created as a result of employing robots? Yes
Title: 1-Adv. Technology(MIS), 1-Mfg. Engr.

45. How many departments were eliminated or consolidated?
No.
Eliminated (identify) ____
Consolidated (identify) ____

46. How has the use of robots affected the number of managers required? (number)
decreased by ____ increased by __3__ no change ____

47. Have new robot-specific management positions been created? (Please specify) Yes, two new lead operators and the mfg. engr.

48. If yes, how were the new positions staffed?
__X__retraining present managers
____promotion of workers ____new hires

49. To your knowledge have there been any personal problems caused by the introduction of robots? No

50. Does your firm have a counselor to deal with worker problems caused by robots? (Full time staff member No, part time No)
Comments:

VII. Future Plans Concerning Robots

51. Do you plan to employ additional robots during the next five years? ___Yes___
___new plant, ___existing plant, _X_both

52. If yes to the above question, has the number of use of robots been decided? _No_____
Please list types of robots and reasons for future use of robots as well as anticipated installation dates.

COMMENTS: Justification of robotics workcells is not a problem due to the tremendous quality advantage, and the relentlessness of the pacing of the robot. These two aspects, which are intangible aspects, do not appear to relate to justification. Financing is not a problem, because robotics are financed like any other computer-like asset. Justification of the purchase is another matter.

The technology requirement in the assembly process dictated that it be done by a machine instead of by a human. Financial justification was not a strong motivator here because we would not have achieved satisfactory quality without the robot.

APPENDIX B
Data For Company D

Questionnaire on the Employment of Robotics by the Small Manufacturer

INSTRUCTIONS: This questionnaire is designed to obtain information about your firm's decision to employ robots in the manufacturing processes. The purposes of collecting the information are to help the researcher complete his Ph.D. in Business Administration at the University of Georgia and to provide information to the small business community and the robotics producers about the important role robotics can play in the small manufacturing firm. If desired, the identity of your firm will not be revealed in the data analysis or the publication of the results. Since little data has been collected to date on the use of robots by small manufacturers, your assistance will be a significant contribution. The questionnaire can be completed in about twenty minutes.

I. Company Profile

1. Your company's industry in which robots are used?
   Telecommunications
2. Title of individual completing this questionnaire.
   Vice President-Strategic Programs
3. The annual sales volume of your company for
   1981:______ 1982:______
   1983:______ 1984: $50 to 100 million
4. The number of employees in your firm in
   1981:______ 1982:______
   1983: 2,200 1984: 1,300
5. Which term best describes your firm (place % on line)?
   47 % make-to-stock
   53 % make-to-order
   ____ % other (please specify)______.
6. Type of process ____ Job shop X Continuous process
   X Assembly line.

7. Approximate number of "end items" in master schedule (excluding service parts): 4,000

8. Approximate number of different part numbers: 20,000

9. Number of levels in bill of materials 12

10. The production planning and control system is
 ____ manual X computerized.

II. Types of Robots and Applications

11. In what year did you first introduce robots into your plant?
 1983

12. How many robots were in use in your plant in Jan 1980? ____, Jan 1981? ____, Jan 1982? ____, Jan 1983? 2, Jan 1984? 5, Jan 1985 6.

13. How many robots are now being used in each of the following uses?

Number		Manufacturer/Type	Cost(complete)
___	spot welding		
___	spray painting/ coating		
3	loaders/ unloaders		$75,000 each
3	assembly	Microbot/Alpha	$150,000 total
___	continuous seam welding		
___	material handlers		
___	other (specify)		

If more space is needed, write data below.

14. Robots were placed in a ___ new plant, ___ old plant, X both?

III. Strategic Factors Considered in Employing Robots

15. Who initiated the interest of your firm in robotics?
 Manufacturing Engineers

16. What factors or events initiated the interest?

    Low labor cost of products from the Far East, i.e., we're competing against $1.00-$2.00/hour labor.

17. Who was involved in making the decision to employ robots (position/function)?

    President, CEO, Dir. of Adv. Engr.

18. What were their roles in the decision process?

    Pres.: cost effective return on investment
    CEO: Capital commitment.
    Director, Adv. Mfg: application.

IV. Principal Reasons for Employing Robots

19. Please rank the following items in order of importance as reasons for the use of robots. Most important will rank number 1. Use N/A for any factor which was not considered.

    __1__ to reduce labor costs
    __6__ to ease compliance with OSHA regulations
    __10__ to reduce capital costs
    __2__ to improve product quality
    __3__ to increase output rate
    __5__ to increase the flexibility in changing products, designs, etc.
    __4__ to decrease material waste
    __7__ to relieve workers of tedious and/or dangerous jobs
    __9__ to remedy high labor turnover problems
    __8__ to relieve shortages of skilled workers
    ____ other factors (please list)

V. User Concerns in Adopting Robots

20. How far in advance were operators made aware that robots would be introduced into their plants? ____3____months

21. How were operators made aware that robots would be placed in the plant?

    ____ through labor union __X__ by meeting with management
    ____ were not notified prior to introduction of robot(s)
    __X__ other (specify) Foremen held meetings.

22. Were there objections to employing robots? _X_ yes, ___ no
23. What were the objections?
    Fear of job loss
24. Who objected? ___ managers, _X_ hourly workers
25. How were the objections overcome?
    Communicating the importance of the robots.
26. Please describe the manner by which you financed the purchase of robots for your plant(s).
    Bank loan.
27. Do you feel that financing of capital equipment is a problem that is delaying the introduction of robotics and/or other advanced technology into the small manufacturers plant?
    Yes
28. What recommendations do you have which might enhance the financing of advanced manufacturing technology for the small manufacturer? This can include sources of financing, taxation, government assistance, and so forth.
    Gov't. Assistance
    Investment incentives

VI. Results of Employing Robots

29. Which of the following occurred after introducing robots?
    _X_ reduced labor costs by _50_ % on jobs robotized.
    _X_ eased compliance with OSHA regulations
    _-_ reduced capital costs by _more capital costs_
    _X_ improved product quality, measure of improvement ______
    ___ increased output rate by _____ %
    _X_ decreased material waste by _5_ %
    ___ relieved shortage of skilled labor
    _X_ relieved workers of tedious and/or dangerous jobs
    ___ reduced labor turnover by _____ %
    ___ increased flexibility in changing products, designs, etc.
    Examples of increased flexibility: Less flexibility.

    ___ other results:

30. Can you readily identify costs and savings attributable to using robots? Examples: Yes, three human workers were removed from robotized jobs.

31. The introduction of robots has caused the following work schedule changes.
    A. Average number of shifts per day: ____increased, __X__same, ____decreased. Current number of shifts is __3__, was__3__.
    B. Average length of work shift per day: ____increased, ____decreased, __X__same. Length is __8hr__, was __8hr__.

32. Have quality control specifications been tightened? __No__

33. The number of quality control employees has __X__decreased, ____increased, ____not changed. Current number __0__, previous __1*__.

34. Have robots had any of the following effects on production?
    __No__given ability to manufacture more complex designs.
    __No__increased efficiency in raw materials use. Measure used:

    __No__reduced setup times by ________%
    __No__enabled the production of new products.
    *Operators now perform quality control checks.

35. How many new positions were created as the result of employing robots? __None__

36. How were new positions filled? (Number of workers)
    ____retraining workers ____hiring new workers

37. If positions were eliminated, how many workers were
    ____transferred without retraining, ____same plant, ____another
    ____retrained for a new position within the plant
    ____retrained for a new position in another plant
    ____laid off ____retired early?

38. In retraining efforts, did the union(s) ____choose who would be retrained, ____support your efforts, but not select retrainees, ____oppose your efforts, __X__not applicable.

39. Since the first use of robots, the number of production related accidents has: ____decreased, __X__no change, ____increased.

40. Was it necessary to develop new worker safety measures? __No__

41. How have robots affected the skill levels required in the following areas?
    A. Operators: ____increased skill level, ____decreased, __X__same.
    B. Maintenance: __X__increased, ____decreased, ____same.
    C. Production Scheduling: __X__increased, ____decreased, ____same.
    D. Materials Handling: ____increased, ____decreased, __X__same.
    E. Other (specify): ____________________
    ____increased, ____decreased, ____same.

42. Did the supplier of the robot help with training? ____No____

43. How long was the training period in days for __5__maintenance, __½__production scheduling, __5__operators, ____other (specify) ____________________

44. Have any new organizational units been created as a result of employing robots? __No__
    Title: ____________________

45. How many departments were eliminated or consolidated?
    ___None___
    Eliminated (identify) ____________________
    Consolidated (identify) ____________________

46. How has the use of robots affected the number of managers required? (number)
    decreased by _____ increased by _____ no change __X__

47. Have new robot-specific management positions been created? (Please specify) No

48. If yes, how were the new positions staffed?
    ____retraining present managers
    ____promotion of workers ____new hires

49. To your knowledge have there been any personal problems caused by the introduction of robots? __No__

50. Does your firm have a counselor to deal with worker problems caused by robots? (Full time staff member __No__, part time __No__)
    Comments:

VII. Future Plans Concerning Robots

51. Do you plan to employ additional robots during the next five years? ___Yes___
    ____new plant, _X_existing plant, ____both

52. If yes to the above question, has the number of use of robots been decided? _No_____
    Please list types of robots and reasons for future use of robots as well as anticipated installation dates.

APPENDIX B
Data For Company E

Questionnaire on the Employment of Robotics by the Small Manufacturer

INSTRUCTIONS: This questionnaire is designed to obtain information about your firm's decision to employ robots in the manufacturing processes. The purposes of collecting the information are to help the researcher complete his Ph.D. in Business Administration at the University of Georgia and to provide information to the small business community and the robotics producers about the important role robotics can play in the small manufacturing firm. If desired, the identity of your firm will not be revealed in the data analysis or the publication of the results. Since little data has been collected to date on the use of robots by small manufacturers, your assistance will be a significant contribution. The questionnaire can be completed in about twenty minutes.

I. Company Profile

1. Your company's industry in which robots are used?
   Hand tools, Automotive & Electrical

2. Title of individual completing this questionnaire.
   Project Engineer

3. The annual sales volume of your company for
   (Company would not release this data.)
   1981:______ 1982:______
   1983:______ 1984:______

4. The number of employees in your firm in
   1981:______ 1982:______
   1983: about 300 1984: about 300

5. Which term best describes your firm (place % on line)?
   10 % make-to-stock
   90 % make-to-order
   ____ % other (please specify)______.

6. Type of process _____ Job shop _____ Continuous process __X__ Assembly line.
7. Approximate number of "end items" in master schedule (excluding service parts): ______________
8. Approximate number of different part numbers: ________
9. Number of levels in bill of materials ____________
10. The production planning and control system is _____ manual __X__ computerized.

II. Types of Robots and Applications

11. In what year did you first introduce robots into your plant? __1985__
12. How many robots were in use in your plant in Jan 1980? _____, Jan 1981? __0__, Jan 1982? __0__, Jan 1983? __0__, Jan 1984? __0__.
13. How many robots are now being used in each of the following uses?

Number		Manufacturer/Type	Cost(complete)
____	spot welding	________	________
____	spray painting/ coating	________	________
3	loaders/ unloaders	________	$250,000 total
____	assembly	________	________
____	continuous seam welding	________	________
____	material handlers	________	________
____	other (specify)	________	________

If more space is needed, write data below.

14. Robots were placed in a ____ new plant, __X__ old plant, ____ both?

III. Strategic Factors Considered in Employing Robots

15. Who initiated the interest of your firm in robotics?
    Project Engineer

16. What factors or events initiated the interest?
    The safety factor and better production flow.

17. Who was involved in making the decision to employ robots (position/function)?
    Manager of Engineering, Project Engineer,
    Quality Control Supervisor.

18. What were their roles in the decision process?
    Equal roles, all working together on this project.

IV. Principal Reasons for Employing Robots

19. Please rank the following items in order of importance as reasons for the use of robots. Most important will rank number 1. Use N/A for any factor which was not considered.
    N/A to reduce labor costs
    5 to ease compliance with OSHA regulations
    N/A to reduce capital costs
    2 to improve product quality
    N/A to increase output rate
    3 to increase the flexibility in changing products, designs, etc.
    4 to decrease material waste
    1 to relieve workers of tedious and/or dangerous jobs
    N/A to remedy high labor turnover problems
    6 to relieve shortages of skilled workers
    ____ other factors (please list)

V. User Concerns in Adopting Robots

20. How far in advance were operators made aware that robots would be introduced into their plants? 6 months

21. How were operators made aware that robots would be placed in the plant?
    ____ through labor union X by meeting with management
    ____ were not notified prior to introduction of robot(s)
    ____ other (specify)

22. Were there objections to employing robots? ___ yes, _X_ no
23. What were the objections?

24. Who objected? ___ managers, ___ hourly workers
25. How were the objections overcome?

26. Please describe the manner by which you financed the purchase of robots for your plant(s).
    Financed by parent company.

27. Do you feel that financing of capital equipment is a problem that is delaying the introduction of robotics and/or other advanced technology into the small manufacturers plant?
    I feel that financing of capital equipment is a problem.
28. What recommendations do you have which might enhance the financing of advanced manufacturing technology for the small manufacturer? This can include sources of financing, taxation, government assistance, and so forth.

VI. Results of Employing Robots

29. Which of the following occurred after introducing robots?
    ___reduced labor costs by ______%
    _X_eased compliance with OSHA regulations
    ___reduced capital costs by ____________
    _X_improved product quality, measure of improvement ____________
    ___increased output rate by ______%
    ___decreased material waste by ______%
    _X_relieved shortage of skilled labor
    _X_relieved workers of tedious and/or dangerous jobs
    _X_reduced labor turnover by ______%
    _X_increased flexibility in changing products, designs, etc.
    Examples of increased flexibility:

    ___other results:

30. Can you readily identify costs and savings attributable to using robots? No Examples: We have not attained the desired reliability of robots.

31. The introduction of robots has caused the following work schedule changes.
    A. Average number of shifts per day: ____increased, _X_same, ____decreased. Current number of shifts is ____, was____.
    B. Average length of work shift per day: ____increased, ____decreased, _X_same. Length is _____, was _____.

32. Have quality control specifications been tightened? _Yes_

33. The number of quality control employees has ____decreased, ____increased, _X_not changed. Current number ____, previous ____.

34. Have robots had any of the following effects on production?
    _No_given ability to manufacture more complex designs.
    _No_increased efficiency in raw materials use. Measure used:

    _No_reduced setup times by _______%
    _No_enabled the production of new products.

35. How many new positions were created as the result of employing robots? _None_

36. How were new positions filled? (Number of workers)
    ____retraining workers ____hiring new workers

37. If positions were eliminated, how many workers were
    ____transferred without retraining, ____same plant, ____another
    ____retrained for a new position within the plant
    ____retrained for a new position in another plant
    ____laid off ____retired early?

38. In retraining efforts, did the union(s) ____choose who would be retrained, ____support your efforts, but not select retrainees, ____oppose your efforts, _X_not applicable.

39. Since the first use of robots, the number of production related accidents has: ____decreased, _X_no change, ____increased.

40. Was it necessary to develop new worker safety measures? _No_

41. How have robots affected the skill levels required in the following areas?
    A. Operators: _X_increased skill level, ____decreased, _X_same.
    B. Maintenance: _X_increased, ____decreased, ____same.
    C. Production Scheduling: ____increased, ____decreased, _X_same.
    D. Materials Handling: ____increased, ____decreased, _X_same.
    E. Other (specify): ________________ ____increased, ____decreased, ____same.

42. Did the supplier of the robot help with training? _Yes_

43. How long was the training period in days for _5_maintenance, _0_production scheduling, _5_operators, ____other (specify) ________________

44. Have any new organizational units been created as a result of employing robots? _No_
    Title: ________________

45. How many departments were eliminated or consolidated? _None_
    Eliminated (identify) ________________
    Consolidated (identify) ________________

46. How has the use of robots affected the number of managers required? (number)
    decreased by ____ increased by ____ no change _X_

47. Have new robot-specific management positions been created? (Please specify) No

48. If yes, how were the new positions staffed?
    ____retraining present managers
    ____promotion of workers ____new hires

49. To your knowledge have there been any personal problems caused by the introduction of robots? _No_

50. Does your firm have a counselor to deal with worker problems caused by robots? (Full time staff member _No_, part time _No_)
    Comments: It is unnecessary

VII. Future Plans Concerning Robots

51. Do you plan to employ additional robots during the next five years? No decision has been made
____new plant, ____existing plant, ____both

52. If yes to the above question, has the number of use of robots been decided? ____________________
Please list types of robots and reasons for future use of robots as well as anticipated installation dates.
Further robotics implementation will depend upon results obtained with present systems. (This comment from an interview with the Project Engineer.)

BIBLIOGRAPHY

1. Abair, D., and J. C. Logan, *13th International Symposium on Industrial Robots and Robots 7,* Society of Manufacturing Engineer's Dearborn, MI, 1983.

2. Abernathy, W. J., and Chakravarthy, B. S. "The Federal Initiative in Industrial Innovation: The Automotive Case," *Sloan Management Review,* Summer, 1979, pp. 5-17.

3. Abernathy, W. J., and Utterback, J. M. "Patterns of Industrial Innovation," *Technology Review,* Volume 80, 1978, pp. 40-47.

4. Alford, J. M. "Strategic Decision Making by the Entrepreneur," *Southern Management Association Proceedings 1983,* pp. 335-337.

5. Alford, J. M. "An Integrative Decision Model for Adopting Robotics Technology," *Southern Management Association Proceedings 1984,* pp. 353-357.

6. Avarde, S. "Technological Innovation—Key to Productivity: The Erosion of American Productivity is a Long-Term Structural Phenomenon; To Increase Productivity We Need Corporate and Government Policies for Advancing the Creation and Application on New Technologies," *Research Management,* July 1982, pp. 33-41.

7. Ayres, R. U., and Miller, S. M. *Robotics Applications and Social Implications,* Ballinger Publishing Company, Cambridge, MA, 1983.

8. Bailey, J. R., "Product Design for Robotic Assembly," *Robots 7,* Op. Cit., pp. 11-44 through 11-57.

9. Banks, R. L., and Wheelwright, S. C. "Operations vs Strategy: Trading Tomorrow for Today," *Harvard Business Review,* May-June 1979, pp. 112-120.

10. Beranek, W., and Selby, E. B., Jr. "Accelerated Depreciation and Income Growth," *Journal of American Real Estate and Urban Economics Association,* Spring 1981, pp. 67-71.

11. Binswanger, H., and Ruttar, V. *Induced Innovation: Technology, Institutions and Development,* John Hopkins University Press, Baltimore, 1978.

12. Blaker, M., Ed. *The Politics of Trade: U.S. and Japanese Policymaking for the GATT Negotiations,* East Asian Club, Columbia University, New York, 1978.

13. Bortz, P. I., and Lipman, W. F. "Federal Incentives for Innovation: The Urban Consortium for Technology Initiative," NSF Report RA-771107, National Science Foundation, Washington, DC 1977.

14. Boulton, W. R., *Business Policy: The Art of Strategic Management,* MacMillan Publishing Company, New York, 1984.

15. Buffa, E. S. "Making American Manufacturing Competitive," *California Management Review,* Spring 1984, pp. 29–46.

16. Butler, G. R., and Nelson, T. J. "The Modern Technology Shift and Existing Small Firms," *Wisconsin Small Business Forum,* Spring 1984, pp. 31–39.

17. Carney, T. P. *False Profits: The Decline of Industrial Creativity,* University of Notre Dame, South Bend, 1981.

18. Carroll, G. R. "The Specialist Strategy," *California Management Review,* Spring 1984, pp. 126–137.

19. Chakrabarti, A. K. "The Cross-National Comparison of Patterns of Industrial Innovations (Canada, France, West Germany, Japan, and The United States)," *Columbia Journal of World Business,* Fall 1982, pp. 33–39.

20. Chandler, A. *Strategy and Structure, Chapters in the History of American Enterprise,* MIT Press, Cambridge, MA, 1962.

21. Channon, D. *The Strategy and Structure of British Enterprise,* MIT Press, Cambridge, MA, 1973.

22. Chase, H. C. "Export Expansion and Small Business Productivity," *American Journal of Small Business,* Spring 1984, pp. 21–27.

23. Christopher, R. C. "Mom, Pop and the Robots," *Japan Society Newsletter,* July–August 1983, pp. 2–9.

24. Collins, E. L. *Tax Policy and Innovation: A Synthesis of Evidence, National Science Foundation,* Washington, DC, 1981.

25. Cooper, A. C. "Strategic Management: New Ventures and Small Business," in *Strategic Management: A New View of Business Policy and Planning,* Schendel, D. E. and Hofer, C. W. (Ed.) Little Brown, Boston, 1979.

26. Crawford, C. M., and Tellis, K. G. J. "The Technological Innovation Controversy," *Business Horizons,* July–August 1981, pp. 76–88.

27. Daft, R. L. "Learning the Craft of Organizational Research," *Academy of Management Review,* October 1983, pp. 539–546.

28. Data Resources Inc. *Macroeconomics Impacts of Federal Pollution Control Laws: 1978 Assessment,* Council on Environmental Quality, Washington, DC, 1978.

29. Delacroix, J. "Export Strategies for Small American Firms," *California Management Review,* Spring 1984, pp. 138–153.

30. Dodge, K. *Government and Business: Prospects for Partnership,* The University of Texas Press, Austin, 1980.

31. Dorf, R. C. *Robotics and Automated Manufacturing,* Reston Publishing Company, Inc., Reston, VA, 1983.

32. Drucker, P. F. "Entrepreneurial Strategies," *California Management Review,* Winter 1985, pp. 9–25.

33. Easton, N. *Reagan's Squeeze on Small Business: How the Administration Plan Will Increase Economic Concentration,* Presidential Accountability Group, Washington, DC, 1981.

34. *The New Industrial Countries and Their Impact on Western Manufacturing,* EIU Special Report Number 73, London, January 1980.

35. Engleberger, J. F. *Robots in Practice,* AMACOM, New York, 1980.

36. Feinman, S., and Fuentevilla, W. *Indicators of International Trends in Innovation,* National Science Foundation, Washington, DC, 1976.

37. Fitch, L. C. "Promoting Technological Innovation in the United States: The Role of Public and Private Sectors," *Federal Incentives for Innovation,* NSF/DRI Task II-14, Washington, DC, 1975, 32 pages.

38. Fleck, J. "The Adoption of Robots," *13th International Symposium on Industrial Robots and Robots 7 Proceedings,* Society of Manufacturing Engineers, Dearborn, MI, 1983, pp. 1.41–1.51.

39. Fredrickson, J. W. "Strategic Process Research: Questions and Recommendations," *Academy of Management Review,* October 1983, pp. 565–575.

40. Froelich, L. "Robots to the Rescue," *Datamation,* January 1981, pp. 85–96.

41. Frohman, A. L. "Putting Technology into Strategic Planning," *California Management Review,* Winter 1985, pp. 48–59.

42. Fromm, G. (Ed.) *Tax Incentives and Capital Spending,* The Brookings Institution, Washington, DC, 1971.

43. Foulkes, F. K., and Hirsch, J. L. "People Make Robots Work," *Harvard Business Review,* January-February 1984, pp. 94-102.

44. Gale, G. T., and Klavan, S. R. "Formulating A Quality Improvement Strategy," *The Journal of Business Strategy,* Winter 1985, pp. 21-32.

45. Gilmore, F. "Formulating Strategies in Smaller Companies," *Harvard Business Review,* May-June 1971, pp. 71-81.

46. Girafalco, L. A. "The Dynamics of Technological Change," *The Wharton Magazine,* Fall 1982, pp. 31-37.

47. Gold, B. *Productivity, Technology, and Capital: Economic Analysis, Managerial Strategies, and Government Policies,* Lexington Books, Heath, Canada, 1979.

48. Greenwood, F., and Gupta, J. "Improving Small Business Productivity," *American Journal of Small Business,* April-June 1983, pp. 15-18.

49. Gresser, J. *High Technology and Japanese Industrial Policy: A Strategy for U.S. Policymakers,* Government Printing Office, Washington, DC, 1980.

50. Gustafson, R. E., "Choosing Manufacturing System, Based on Unit Costs," *Robots 7,* Op. Cit., pp. 1-41 through 1-51.

51. Hagedorn, H. J. "The Factory of the Future: What About the People," *The Journal of Business Strategy,* Summer 1984, pp. 38-45.

52. Hami-Noori, A., and Templer, A. "Factors Affecting the Introduction of Industrial Robots," *International Journal of Operations & Production Management,* Volume 3, Number 2, pp. 46-57.

53. Higgins, V. A. "An Idea Whose Time Had Come: Productivity—The International Perspective," *American Journal of Small Business,* Spring 1984, pp. 1-16.

54. Hofer, C. W., and Schendel, D. *Strategy Formulation: Analytical Concepts,* West Publishing Company, St. Paul, 1978.

55. Holloman, H. J. "Government and the Innovation Process," *Technology Review,* May 1979, pp. 30-41.

56. Holloman, H. J. "Technology in the United States: The Options Before Us," *Technology Review,* July-August 1972, pp. 32-42.

57. Horowitz, B., and Kolodny, R. *The Economic Effects of Uniformity in the Financial Reporting of R&D Expenditures,* National Science Foundation, Washington, DC, 1980.

58. Iacocca, L. A. "Making U.S. Business Competitive," *The Journal of Business Strategy,* Summer 1984, pp. 26–28.

59. JMA Research Institute, *Robotization: Its Implications for Management,* Fuji Corporation, Tokyo, 1983.

60. Johnson, C. *Japan's Public Policy Companies,* American Enterprise Institute for Public Policy Research, Washington, DC, 1978.

61. Johnson, C. *MITI and the Japanese Miracle: The Growth of Industrial Policy, 1925–1975,* Stanford University Press, Stanford, CA 1982.

62. Jones, W. D. "The Nature of Strategic Planning in Small Business," *Southern Management Association Proceedings,* 1980, pp. 218–220.

63. Kelly, P. et al. *Technological Innovation: A Critical Review of Current Knowledge,* The San Francisco Press, San Francisco, 1978.

64. Kennedy, E. M. "Revitalizing the U.S. Economy: The Role of Industrial Strategy," *The Journal of Business Strategy,* Summer 1984, pp. 4–8.

65. Kidder, L. H. *Seltiz Wrightman & Cook's Research Methods in Social Relations,* Holt, Rinehart, and Winston, New York, 1981.

66. Kirby, S. *Japan's Role in the 1980s,* EIU, London, 1980.

67. Kirby S. *Towards the Pacific Century: Economic Development in the Pacific Basin,* EIU, London, 1983.

68. Kirchoff, B. A. and Knight, W. E. "Government's Role in Research and Development," *Frontiers of Entrepreneurship,* Proceedings of the 1981 Conference on Entrepreneurship at Babson College, pp. 321–336.

69. Kitti, C. "Patent Invalidity Studies: A Survey," *Ideas,* Volume 20, Number 1, 1979, pp. 55–76.

70. Knod, E. M., Jr. et al. "Robotics Challenges for the Human Resources Manager," *Business Horizons,* March/April 1984, pp. 38–46.

71. Koch, D. L. "The Micro-Solution," *Economic Review.* Federal Reserve Bank of Atlanta, September 1983, pp. 34–41.

72. Lambrinos, J., and Johnson, W. G. "Robots to Reduce the High Cost of Illness and Injury," *Harvard Business Review,* May–June 1984, pp. 24–28.

73. Lawrence, P. R., and Dyer, D. *Renewing American Industry,* The Free Press, New York, 1983.

74. Lederman, L. L. "Federal Policies and Practices Related to R&D/Innovation," *Research Management,* May 1978, pp. 18–20.

75. Legler, J. B., and Hoy, F. "Productivity in the Small Business Sector, 1965–1976," *American Journal of Small Business,* April–June 1984, pp. 49–55.

76. Leontiades, J. "Market Share and Corporate Strategy in International Industry," *The Journal of Business Strategy,* Summer 1984, pp. 30–38.

77. Levitt, A., Jr. "Industrial Policy: Slogan or Solution?" *Harvard Business Review,* March/April 1984, pp. 6–8.

78. Lippit, J. W., and Oliver, B. L. "The Productive Efficiency and Employment Implications of the SBA's Definition of Small," *American Journal of Small Business,* January–March 1984.

79. Logsdon, T. *The Robot Revolution,* Simon & Schuster, Inc., New York, 1984.

80. MacMillian, I. C. "Strategy and Flexibility in the Smaller Business," *Long Range Planning,* Volume 8, Number 3, 1975, pp. 62–63.

81. Magaziner, I. C., and Hout, T. M. *Japanese Industrial Policy,* Institute of International Studies, University of California, Berkeley, 1981.

82. Mansfield, E. "Contribution of R&D to Economic Growth in the United States," *Science,* February 4, 1972, pp. 477–486.

83. Marsh, F. *Japanese Overseas Investment,* EIU, London, 1983.

84. McCullough, R. "Trade Deficits, Industrial Competitiveness and the Japanese," *California Management Review,* Winter 1985, pp. 140–156.

85. McKenna, J. F., and Oritt, P. L. "Growth Decision Making for the Small Firm," *Southern Management Association Proceedings,* 1980, pp. 221–223.

86. McLean, M. (Ed.) *The Japanese Electronic Challenge,* Technova, St. Martin's Press, New York, 1982.

87. Mensch, G. *Stalemate in Technology,* Ballinger Publishing Company, New York, 1979.

88. Mensch, G., and Newhaus, R. J. (Eds.) *Work, Organization, and Technical Change,* Plenum Press, New York, 1982.

89. Mintzberg, H., Raisinghani, D. and Theoret, A. "The Structure of 'Unstructured' Decision Processes," *Administrative Science Quarterly,* Volume 21, pp. 246-275.

90. Morey, N. C., and Luthans, F. "An Emic Perspective and Ethnoscience Methods for Organizational Research," *Academy of Management Review,* January 1984, pp. 27-36.

91. Noll, R. G. *Government Policies and Technologies Innovation: Project Summary,* National Science Foundation, Washington, DC, 1975.

92. Office of Technology Assessment, *Automation and the Work-Selected Labor, Education, and Training Issues,* Washington, DC, 1983.

93. Okun, A. M., and Perry, G. L. (Eds.) *Brookings Papers on Economic Activity,* The Brookings Institute, Washington, DC, 1978.

94. Organization for Economic Cooperation and Development, *The Industrial Policy of Japan,* Paris, 1972.

95. Oulette, R. P. et al. *Automation Impacts on Industry,* Ann Arbor Science Publishers, Ann Arbor, MI, 1983.

96. Owen, A. E. *Chips in Industry,* EIU, London, 1982.

97. Porter, M. E. *Competitive Strategy,* The Free Press, New York, 1980.

98. Porter, M. E. "Technology and Competitive Advantage," *The Journal of Business Strategy,* Winter 1985, pp. 60-78.

99. Reich, R. B. "What Kind of Industrial Policy," *The Journal of Business Strategy,* Summer 1984, pp. 10-17.

100. *Research Management,* Special Issue on National Technology Policy, January 1977.

101. Rice, G. H., and Hamilton, R. E. "Decision Theory and the Small Businessman," *American Journal of Small Business,* Fall 1979, pp. 7-15.

102. Robinson, R. B., Jr. "Small Firm Performance in the 1980s: Will Planning Contribute or Detract," *Southern Management Association Proceedings,* 1980, pp. 25-27.

103. Robinson, R. B., Jr., and Freeman, B. B. "The Impact of SBDC Consulting on the Performance of New and Existing Small Businesses," *Southern Management Association Proceedings,* 1983, pp. 330-332.

104. Robinson, R. B., Jr., and Pearce, J. A., II. "Research Thrusts in Small Firm Strategic Planning," *Academy of Management Review,* January 1984, pp. 128–137.

105. Robinson, R. B., and Pearce, J. A., II. "Impact of Formalized Strategic Planning on Financial Performance in Small Organizations," *Strategic Management Journal,* Volume 4, Number 3, 1983, pp. 78–87.

106. Robot Institute of America, "The Decline of Productivity and the Resultant Loss of U.S. World Economic and Political Leadership," RIA Position Paper, Dearborn, MI, 1982.

107. Robot Institute of America, News Release, Dearborn, MI, December 12, 1984.

108. "Robotics: Issues and Trends Briefing Report," Edison Electric Institute, Washington, DC, 1983.

109. Roosa, R. V., Matsukawa, M., and Gutowski, A. *East–West Trade at a Crossroads,* New York University Press, New York, 1982.

110. Rosato, P. J., "Robotic Implementation—Do It Right," *Robots 7,* Op. Cit., pp. 4-32—4-49.

111. Rumelt, R. *Strategy, Structure and Economic Performance,* Harvard University Press, Cambridge, MA, 1974.

112. Sahal, D. *Patterns of Technological Innovation,* Addison-Wesley Publishing Company, Inc., Reading, MA, 1981.

113. Saso, M., and Kirby, S. *Japanese Industrial Competition to 1990,* ABT Associates, Cambridge, MA, 1982.

114. Schiffel, D., and Kitti, C. "Rates of Invention: International Patent Comparisons," *Research Policy,* Volume 7, 1978, pp. 324-340.

115. Schoneberger, R. J. *Japanese Manufacturing Techniques,* The Free Press, New York, 1982.

116. Scott, B. R. "National Strategy for Stronger U.S. Competitiveness," *Harvard Business Review,* March–April 1984, pp. 77–91.

117. Shrivasta, P., and Mitroff, I. I. "Enhancing Organizational Research Utilization," *Academy of Management Review,* January 1984, pp. 18–26.

118. Skinner, W. "Manufacturing—Missing Link in Corporate Strategy," *Harvard Business Review,* May–June 1969, pp. 136–145.

119. *The State of Small Business; A Report to the President,* Small Business Administration, U.S. Government Printing Office, Washington, DC, March 1984.

120. Soergel, D. G. "High-Tech Freeze-Out: An R&D Specialist Tells How Government Policies Make Life Tough for the Independent Innovator," *Reason,* April 1982, pp. 40–43.

121. Spilhaus, A. "Miniature Mechanical Marvels," *Technology Review,* January 1982, pp. 51–57.

122. "The State of the U.S. Machine Tool Industry: U.S. Department of Commerce Briefing," May 6, 1983, National Machine Tool Builders Association, Washington, DC, 1983.

123. Suzaki, K. "Japanese Manufacturing Techniques: Their Importance to U.S. Manufacturing," *The Journal of Business Strategy,* Winter 1985, pp. 10–19.

124. Tanner, W. *Industrial Robots,* Society of Manufacturing Engineers, Dearborn, MI, 1978.

125. Thomas, K. W., and Tymon, W. G., Jr. "Necessary Properties of Relevant Research: Lessons from Recent Criticisms of the Organizational Sciences," *Academy of Management Review,* July 1982, pp. 345–352.

126. Thurston, P. H. "Should Smaller Companies Make Formal Plans?" *Harvard Business Review,* September–October 1983, pp. 162–184.

127. Topperwein, L. L. et al. *Robotics Applications for Industry: A Practical Guide,* Noyes Data Corporation, Park Ridge, NJ, 1983.

128. Tornatzky, L. G. et al. *The Process of Technological Innovation: Reviewing the Literature,* National Science Foundation, Washington, DC, 1983, p. 190.

129. *Competitive Position of U.S. Producers of Robotics in Domestic and World Markets,* United States International Trade Commission, Washington, DC, 1983.

130. Utterback, J. "Innovation in Industry and the Diffusion of Technology," *Science,* February 15, 1974, pp. 620–626.

131. Uyehava, C. H. (Ed.) *Technological Exchange: The U.S.—Japanese Experience,* University Press of America, Inc., Washington, DC, 1982.

132. Van Blois, J. P. "Economic Models: The Future of Robotic Justification," *Robots 7,* pp. 4.24–4.31.

133. Visscher, M. *Tax Policies for R&D and Technical Innovation,* National Science Foundation, Washington, DC, 1976.

134. Warnecke, H. J., and Schraft, R. D. (Eds.) *Industrial Robots: Application Experience,* I.F.S. Publications, Ltd., Bedford, England, 1982.

135. Whaley, G. "The Impact of Robotics Technology Upon Human Resource Management," *Personnel Administrator,* September 1982.

136. Wheelwright, S. C., *Harvard Business Review,* July–August 1981, p. 67–74.

137. Wiesner, J. "Has the U.S. Lost Its Initiative in Technological Innovation?" *Technology Review,* July/August 1976, pp. 54–60.

138. Woo, C. "The Surprising Case for Low Market Share," *Harvard Business Review,* November–December 1982, pp. 106–113.

INDEX

Names

Subjects

One
Week
Loan